The Institute of Biology's
Studies in Biology no. 148

Cell Growth
and Division

Denys N. Wheatley
Ph.D., D.Sc., M.R.C.Path., F.I.Biol.
Reader in Cell Pathology,
University of Aberdeen

Edward Arnold

First published 1982
by Edward Arnold (Publishers) Limited
41 Bedford Square, London WC1 3DQ

British Library Cataloguing in Publication Data

Wheatley, Denys N.
  Cell growth and division.—(Studies in biology; no. 148)
  1. Cell proliferation   2. Cell cycle
  I. Title   II. Series
  574.87′61   QH605

  ISBN 0–7131–2859–3

Photoset and printed by Photobooks (Bristol) Ltd

BST

General Preface to the Series

Because it is no longer possible for one textbook to cover the whole field of biology while remaining sufficiently up to date, the Institute of Biology proposed this series so that teachers and students can learn about significant developments. The enthusiastic acceptance of 'Studies in Biology' shows that the books are providing authoritative views of biological topics.

The features of the series include the attention given to methods, the selected list of books for further reading and, wherever possible, suggestions for practical work.

Readers' comments will be welcomed by the Education Officer of the Institute.

1982

Institute of Biology
41 Queen's Gate
London SW7 5HU

Preface

The processes of cell growth and division are of fundamental importance to biology, yet our understanding of them is almost as superficial as it was nearly one hundred years ago when E. B. Wilson first published his magnificent work *The Cell in Development and Heredity* (1896 Macmillan, New York), a book worth consulting today. Many simple questions can be asked – how does a cell 'know' when to divide, and how does it manage to do so into two daughters of the same size? But simple questions do not necessarily have simple answers, as any cell biologist knows.

The intention of this small book is not only to deal with some of the straightforward detail about cell growth and division, but to give some idea as to how some of the problems are being tackled. Intuitively one feels that a subject has been 'grasped' when intelligent questions can be asked about it. This is probably the best attitude to adopt here, as in approaching any scientific problem.

Studies in Biology No. 21 on *Cell Division and Heredity* deals with the segregation of genes to progeny; it did not deal with processes of preparation for division, nor with mitosis. Thus there is no overlap and the two books are entirely complementary.

Aberdeen, 1982

D. N. W.

Contents

1 Introduction

1.1 Unity in diversity

There is an enormous amount of diversity in nature, and much diversity in the cells which make it up. To understand how cells grow and divide, scientists have had to select representatives from many of the major groups of living organisms, some for superficial study, others for in-depth analysis. The latter group includes such celebrated species as the bacterium *Escherichia coli*, the ciliated protozoan known as *Tetrahymena pyriformis*, a division yeast called *Schizosaccharomyces pombe*, *Allium* (onion) meristem cells, a fungus named *Physarum polycephalum*, sea urchin eggs and mouse fibroblasts (e.g. 3T3 cells). In a short book only a glimpse can be given at what these organisms alone have taught us, for each exemplifies some aspect of cell growth and division. We must therefore look for underlying principles. There is no sound theory of cell proliferation, and formulation of one would rely heavily on common features found in comparisons between many different types of cell. But is there any reason to suppose that common principles are at work among diverse organisms?

Broadly speaking, cells have a far greater similarity than diversity when one looks at their organelles or some of their proteins and nucleic acids. It is largely the way these macromolecules are assembled which accounts for the outward diversity of form. At the molecular level, the cytoplasm of onion root cells contains materials which are difficult to distinguish from those found in sea urchin eggs. When life emerged on this planet, it was fashioned by the chemistry of carbon, along with nitrogen, oxygen, hydrogen and a handful of other elements in smaller quantities. Earth scientists recognize four universal attributes of living things: proteins, nucleic acids, cells and membranes. In physiological terms, they recognize three ubiquitous functions: cell growth, cell division and self-replication. However much these arbitrary criteria may be criticised by biologists, it would be difficult to deny that cell growth and division are 'primitive' in nature. The greater the antiquity, the more probable it is that a common mechanism is involved. Evolution has simply produced variations upon this theme.

The process of cell division is intimately concerned with the segregation of chromosomes, the storage modules for the genetic information, which must be passed on from generation to generation. In *Studies in Biology* No 21, Dr Kemp dealt almost exclusively with the importance and strict requirements for the mixing and segregation of genes at division, as seen by the geneticist at meiosis. There is a ring of familiarity about the notion that the cell is the genes' mechanism of perpetuating themselves (DAWKINS, 1978). The reverse would be to argue that the genes are the cell's means of perpetuating itself. Whether we approach the problem from the viewpoint of the gene or the cell, the process of cell division involves segregating both the cytoplasm and the nuclear contents.

The preparations made by a cell for division should not be underestimated. In general, when cells of a metazoan animal are busy doing one job, e.g. making mucus or sending electrical impulses, they cannot be dividing – this is often called the 'division of labour'. When cells divide, they have to put nearly all their effort and substance into it, leaving little for other specialized functions. In cells which are incessantly dividing, there is virtually no trace of any differentiated function; they have been called undifferentiated or dedifferentiated cells, probably because the process of cell division is so basic that it is rarely thought of as a differentiated state itself.

In cells which are continuously dividing, growth occurs between successive mitoses in a period called interphase. It represents a period almost entirely devoted to active preparation for the next division. An incorrect impression about proliferating cells is that preparations for division occur in the last few minutes before an impending division. The well-prepared cell goes into division in a co-ordinated manner and each of the offspring to emerge should have roughly (but not necessarily exactly) the same amount of inheritance and an equal chance of survival. If the preparations for division are disturbed, all kinds of aberrations may occur at division leading to grossly abnormal or non-viable offspring.

Daniel Mazia, one of the foremost exponents on cell division, considers the problem of ensuring the accurate division of the cell as one of the most fundamental of biological problems (DIRKSEN et al., 1978). The discovery of how a cell divides remains an enormous challenge to the scientist. And scientist is used here intentionally rather than biologist because those investigating the problem today include biophysicists, crystallographers, geneticists, organic chemists, molecular biologists and many others. It is a truly interdisciplinary area of research, and its discoveries have impact on all biological and medical sciences.

1.2 Growth and entropy

Some questions about cells have no easy answer, for example, why do cells grow? In speculating about the features of the earliest living organisms in evolution and knowing how 'primitive' organisms behave today, it seems clear that they must not only abide by the same laws of thermodynamics as everything else but that they have been limited by carbon (organic) chemistry. In a system in which polymerization is a major feature, molecular randomness is lost, i.e. entropy decreases as more orderly structures evolve. Under these conditions, rapid development and evolution must have continued because the opposite would otherwise have occurred and any developing system would therefore have reverted to complete chaos (maximum entropy). There is no option for a system with decreasing entropy but to evolve. With the passage of time, those systems which were best suited to continuing to reduce the entropy persisted while other systems (species) disappeared. The adaptability of a protein-based life form combined with mutable nucleic acids saw the evolution of many regulatory controls which have allowed cells to grow and divide, and to colonise many extreme and bizarre niches of this planet.

1.3   Self-assembly

One important concept which needs to be fully appreciated is that cells grow by a self-assembly process. Unlike a sweater or a house, the cell has first to obtain its raw material (nutrients), synthesize its own building units (macromolecules), and then construct itself from the inside out without any reference to some pattern or external plan. There is no schedule at hand to consult as to when events should occur in relation to one another, and there is no timekeeper or quantity surveyor to give the go ahead for division. The cell does not have a balance with which to ensure that its daughters will be the same size at division. Too often we take all these processes for granted because of our familiarity with growth and division, but it is important to contemplate on how they all take place. It has often been presumed that the cell is programmed, i.e. acts in a deterministic manner at the molecular level to make the right things happen within the cell. But there is nothing magical about the behaviour of the molecules within a cell. Growth is achieved through a remarkable process of self-assembly steps within an otherwise random system. This may be difficult to understand but a simple example will serve to show how assembly and regulation may be achieved in the growth of a cell without external control. Imagine that a cell grows cytoplasmic processes just 6 μm in length. What regulates this growth so that it stops at 6 μm? In Fig. 1–1 a possible mechanism is given in molecular terms based on a very simple principle, in which the subunits have only one way of linking together and need no external regulation to ensure accurate length. This is, of course, hypothetical but similar processes will be met in Chapter 4.

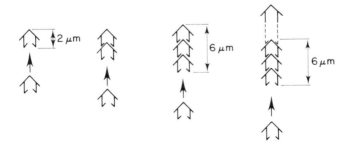

Fig. 1–1   Growth to a constant length of 6 microns. As a fourth subunit joins the chain, the bonds of the first are weakened and the subunit is lost.

1.4   The idealized cell

There are two main classes of organism occurring in nature, *prokaryotes* and *eukaryotes*. Prokaryotic cells do not have their chromosomal material enclosed within a membrane-bound nucleus whereas eukaryotic cells do. Not a great deal of difference is found in their growth processes but division is different, with most eukaryotic cells losing their nuclear membranes for a short period during chromosome segregation. Fig. 1-2 shows a cell of each type in an idealized simplified diagrammatic form. Both have cell coats, cell membranes, cytoplasm containing ribosomes and many other structures.

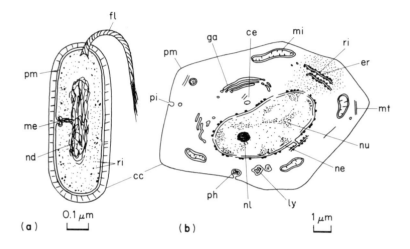

Fig. 1-2   Diagram of a rod-shaped, prokaryotic bacterium (**a**) and an eukaryotic animal cell (**b**). cc, cell coat; ce, centrioles in cell centre; er, endoplasmic reticulum; fl, flagellum; ga, Golgi apparatus; ly, lysosomes; me, mesosome; mi, mitochondrion; mt, microtubule; nd, nucleoid; nl, nucleolus; nm, nuclear membrane; nu, nucleus; ph, phagocytic vacuole; pi, pinocytotic vacuole; pm, plasma membrane; ri, plasma membrane.

Most prokaryotes are *autotrophs*, organisms which can synthesize their food from simple carbon and nitrogen sources. Plant cells (eukaryotes) are also autotrophs whereas animal cells (eukaryotes) are not. The latter require more complex precursors or nutrients and are known as *heterotrophs*, being ultimately dependent upon autotrophs for organic precursors. All organisms are also dependent upon energy to assimilate materials into their biomass and this comes directly or indirectly from the sun. The efficient conversion of solar energy into chemical energy has allowed the successful colonization and adaptation of green plants throughout this planet.

1.5 The control of growth

There are three aspects to be looked at in the control of growth. The first concerns the intrinsic factors which influence the rate at which biomass increases according to the availability of a supply of food or raw material and also to the environmental conditions (Chapters 2 and 3). The second is the mechanism which initiates and co-ordinates division (Chapters 4 and 6). And the third deals with the control of cell numbers in the populations of cells which arise from division (Chapters 2 and 7). All three aspects must be carefully co-ordinated but the mechanisms by which this is achieved remain relatively obscure.

1.6 Degeneration and death

Although growth is concerned with the increase of units of biomass, it is not without its important counterpart, cell death. If a bacterium such as *E. coli* is dropped into a nutrient broth, it will multiply rapidly until it has exhausted all assimilable substances. Since life requires that a continuous supply of substrates for energy generation be available, the effect of uncontrolled proliferation in the organisms would be the cutting off of their own life-line. To avoid this consequence, cells usually slow down their growth when conditions become overcrowded and start to deteriorate. Bacteria often sporulate. This reduces to a

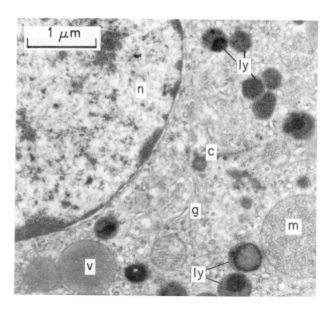

Fig. 1–3 Electron micrograph of an adrenal cell with many lysosomes (ly). Also seen are the nucleus (n), mitochondrion (m), lipid vacuole (v), Golgi apparatus (g) and a centriole (c).

minimum the amount of energy required to support life, at least for some of the cells which can survive until favourable conditions return. Many of the cells, however, will die off during the process. In all cultures this process of drop-out can be occurring at a much less significant level. If growth is slower, and drop-out is occurring, there will be a point at which the two processes balance – a 'steady state'. While our primary concern is with cell growth, much has to be learned about degradation, degeneration, atrophy and death in biological systems. These have been much less popular areas of research and our knowledge is only fragmentary. Sometimes, cell death can be seen as an active process, usually involving organelles called *lysosomes* (Figs. 1–2 & 1–3), which behave like suicide capsules in the cytoplasm. On other occasions, cells undergo a protracted involution by a process called apoptosis, which describes their being shed like falling leaves. The relationship of cell death to the regulation of cell populations will be discussed in section 7.7.

2 Cell Growth and Multiplication

2.1 Growth and cell size

2.1.1 General features

Cell size can increase in several ways, the most usual being by the synthesis and accumulation of new macromolecules and their attendant substances so that the consistency and specific gravity remain constant. Although this seems self-evident, increase can also occur by the accumulation of water. Absorption of water can be a regulated process up to a certain level but if it goes too far, the cell membrane becomes extended and blebbing occurs. Finally the cell bursts and its contents come spewing out (lysis). Oedematous change, as transient wateriness is called, is reversible in tissues which have been injured. Although cells in this condition obviously get larger, this can scarcely be called cell growth. For similar reasons, we might also consider that adipocytes (fat cells), which store large quantities of lipid in intracellular vacuoles, or plant cells, in which a large central vacuole filled with watery sap produces marked elongation, 'grow' in the broadest sense of the word. Clearly 'growth' requires qualification in different circumstances, and when used by biologists without further qualification it generally implies a *co-ordinated accumulation of cellular components with time*. It also implies a simultaneous and equivalent increase in functional capacity.

With the cell as the basic unit of growth, an increase in size usually leads to division and an increase in cell number; this relationship is one of the central issues in biology. Cells of a particular type have characteristic features by which they are recognized, one of which is size. This alone helps us distinguish one species of bacillus from another, or platelets from erythrocytes in the blood.

2.1.2 Cell weight and composition

Weight (mass) is an obvious means of measuring growth. Tissues or organisms weighed in their natural state are measured in grammes as a *wet weight*, often expressed in milligrammes (mg; 10^{-3}g), microgrammes (µg; 10^{-6}g), nanogrammes (ng; 10^{-9}g) or even picogrammes (pg; 10^{-12}g) because cells are very small. Evaporating off water gives the *dry weight*, which is usually expressed as an absolute weight relative to a known wet weight from which it was derived. Sometimes the dry weights of similar organs may be closer than their wet weights, which could simply reflect hydration differences. An example of this is the endometrium or lining of the uterus, which shows considerable changes in the degree of hydration of the cells during the menstrual cycle.

Dry weight can be further analysed by combustion or ashing of the material.

This drives off many of the more volatile products of elements such as carbon, hydrogen, oxygen and nitrogen of the macromolecules and leaves salts and oxides of other elements, e.g. K, Na, Ca, Zn, Cu, Mg, of which the rarer ones are called *trace elements* (e.g. Zn, Co) because of their exceptionally low levels in living materials. Nevertheless, they are vital to metabolism, and growth cannot occur in their absence. Ashing concentrates these substances, making assays more accurate.

Although it is easy to weigh a growing baby or plant, it is not so easy to weigh a cell; and it is also of dubious practical value. But the cell has been weighed by applying the Cartesian diver principle. The late Professor Zeuthen loaded individual amoebae on finely-drawn capillary tubes, giving them buoyancy in a column of water by the inclusion of a small bubble of air. By applying hydrostatic pressure, the diver could be kept at a definite mark. The difference in pressure required to keep the unloaded capillary at the calibration mark could be expressed as a weight difference down to 10 pg.

An alternative to the slow, technically-demanding diver method is to use *interference microscopy.*. This requires a special microscope designed to send one cone of light from the condenser through the specimen, where the macromolecules in cells refract the light to different degrees. The waves coming to the eye-piece are retarded to a greater or lesser extent compared with those of an outer cone of light not passing through the specimen. Interference occurs when the two beams of white light are re-converged and a distinctive coloration similar to the effect of petrol spread on water is produced. Using monochromatic light, it is possible to measure this interference since it is proportional to both the thickness of the object refracting the light and the density of proteins and other macromolecules which are causing this refraction within the specimen. The technique is very sensitive and can be applied at the subcellular level to measure organelles.

But without these specialized techniques, the easiest way of deriving an approximate cell weight is simply by weighing a large, known number of cells, and calculating an average value per cell.

Cells consist of about 70–75% water and 16–20% protein, with the remaining 7–10% including DNA, RNA, lipid, sugars, polysaccharides, vitamins, ions, trace elements, etc. The predominance of protein makes it a useful guide to overall biomass and is often used to monitor the growth characteristics of living organisms.

2.1.3   Analysis of macromolecular increase during growth

Chemical and biochemical assays   Straightforward chemical and biochemical procedures can be used to study macromolecular changes during cell growth. The limitation here is sensitivity because many of the more traditional procedures are too insensitive for assays on small numbers of cells. In general a plentiful supply of material is required, and results are usually expressed as average amounts per unit weight or number of cells. There is complete loss of data at the subcellular level unless fractionation of cellular components is carried out by *differential centrifugation*. But organelles in different fractions often acquire or lose materials during preparation and therefore estimates are not entirely reliable by this technique.

Standard procedures for protein include the celebrated Lowry test for protein which gives a blue coloration when first Cu^{2+} and then a molybdotungstate complex are reacted with aromatic residues in protein. Other similar tests such as the orcinol or diphenylamine reactions allow one to assay RNA and DNA respectively.

But much more sophisticated analytical procedures are now in use following extraction of the specific type of macromolecule of interest, e.g. lipid or polysaccharide. Using, for example, gas chromatography, ion exchange resins, automatic amino acid analysers, protein sequencers, and immunoassay procedures, characteristic groups or regions within major macromolecules can be clearly identified and usually quantified.

Enzymes are macromolecules which readily lend themselves to assay because they react with specific substrates. By formation of an identifiable product from a known concentration of substrate, reaction rates can be estimated. The rate of reaction under standard incubation conditions is, within limits, proportional to the amount of enzyme present. In fact, the estimate is only a rough guide to the number of enzyme molecules in an active state at the time of incubation. Unfortunately, biological material is usually homogenized before assay which means that a true value corresponding to the native enzymes in the original cells is unlikely to be obtained because many previously active molecules are denatured and inactivated. This can be avoided to a certain extent by carrying out the reaction *in situ* at the cellular level, as described below.

Single cell analysis This relies on extremely sensitive techniques for detecting small amounts of macromolecules, by exploiting some characteristic grouping with a unique absorption spectrum, by attaching a fluorescent substance ('dye') to them, or by some similar procedure. Coupled with a scanner, the *fluorescence microscope* and the *integrating microdensitometer* can, for example, measure emitted light of different wavelengths from fluorescent labels which tag on to certain molecules, thereby quantifying the amount per cell. Obviously the best results are obtained when the tag titrates specifically with a known macromolecular species. Use of fluorescent tagged antibodies now makes *immunoassay techniques* potentially capable of measuring with a high degree of accuracy many of the cellular proteins that occur at extremely low concentrations within the cell. Progress in electronics has also led to automation of fluorescent-binding assays. For example, it is possible to measure the amount of DNA in a living cell by exposing it to a substance such as mithramycin which titrates specifically with DNA. It then passes through a beam of UV light and the bound mithramycin fluoresces, the emitted light being picked up by a photosensitive cell and the signal amplified. The process allows a whole population of cells to be analysed in a very short space of time, using an apparatus called a *flow cytofluorimeter* (see also Chapter 3).

Spectrometric methods are still of great use in studying enzymes at the cellular level. Provided a specific enzyme action can be carried out yielding either a coloured product or a product which can be identified in some other way, then a method is available for quantifying the amount of enzyme per cell. *Cytochemistry* demonstrates an important point. Two cells lying side by side in the same tissue may look alike but one cell can frequently show intense activity for an enzyme whereas its neighbour shows none. It can also give intracellular location with considerable accuracy in many cases, e.g. acid phosphatase in the Golgi apparatus and lysosomes. Thus the spatial distribution of enzymes within tissues can be resolved whereas biochemical techniques provide no such information.

Radioactive precursor substances are particularly useful at the single cell level when they become incorporated into macromolecules being synthesized. Treated cells are washed, fixed and overlaid with a photographic film or emulsion, which visualizes the sites that have taken up radioactivity (see Fig. 2–1). *Autoradiography*, as this process is called, and cytochemistry have both been applied at the electron microscope level to give high resolution to the localization of substances.

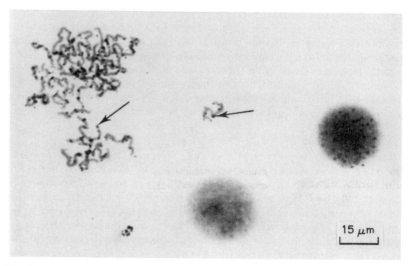

Fig. 2–1 Autoradiograph of 3 nuclei from human lymphocytes stimulated with the lecten, phytohaemagglutinin. One labelled nucleus on right; below, unlabelled nucleus. The third is in mitosis, with grains on individual chromosomes (arrows). By courtesy of Professor J. Swanson Beck.

2.1.4   Comparative aspects of macromolecular synthesis

The accretion of new macromolecules in growing cells is a highly organized affair, not just an indiscriminate accumulation of more molecules. Obviously it would be preferable to gauge increase by an internal yardstick in addition to having it recorded on a per cell basis. The DNA content of a cell is often used in this capacity because its amount per cell remains more or less constant. Thus the alteration of, for example, sphingomyelin, mRNA or arginine kinase in growing cells is often expressed as an amount per unit of DNA and therefore allows each to be independently related to one another.

The fact that DNA content per cell is constant, and each cell type possesses a characteristic amount, suggests that it determines exactly how big a cell will grow. But DNA content is not that invariable. Apart from the content doubling during each cell cycle (Chapter 3), maturation and differentiation of cells from a growing stock sometimes leads to various types of abortive cell division. For example, DNA may double in content and the chromosomes can become segregated within the cell, but the rest of division fails, leaving the cell with twice its normal complement. The usual *complement* of DNA is 2C, but before division this increases to 4C and the two daughters therefore return to the 2C level. The 'divided' cell which has undergone the poorly understood process of *endomitosis* remains at 4C amount of DNA. The process can be repeated several times giving cells of 8, 16, 32 and higher complement numbers or 'ploidy'. *Polyploidization* is particularly frequent in such places as liver cells, many cultivated plants and large protozoa, but it is a widespread phenomenon and occurs to some degree in most eukaryotic cells. Ploidy refers to the number of chromosome sets possessed by cells; haploid is a half-set (n) equivalent to C DNA, diploid is 2n (equivalent to 2C), tetraploid 4n (equivalent to 4C). You may wonder why two expressions have come into use. Ploidy is a biological expression relating to numbers of sets of chromosomes (n) whereas complement (C) refers to a biochemical measure of DNA content. While they generally agree, it is possible for these values to be disparate.

Rise in ploidy class is accompanied by an increase in cell mass because cytoplasmic division fails to follow the doubling of all the essential components. Clearly a hexadecaploid (16C) cell is much larger than a diploid (2C) cell, but not necessarily 8 times bigger. The size of the cell could be dictated by the number of chromosomes, since everything is ultimately dependent upon the genes they carry. This idea is by no means new; Hertwig in 1908 had already commented that the size of a cell seemed to be closely related to the size of its nucleus, and after seventy years, we seem little further advanced in understanding the basis of this relationship. Indeed the relationships between (a) nuclear and cell size, and (b) cell size and the initiation of division, present two of the most basic and challenging questions in biology today.

2.2 Cell number

2.2.1 Cell counting and sizing

Since increase in cell size and number occurs in growing populations, accurate methods of measuring these parameters are essential. The laborious method of counting cells in a haemocytometer (see Fig. 2–2) has been largely superseded by electronic methods. Nevertheless, the haemocytometer is inexpensive, allows the investigator to see the cells being counted, requires a very small sample and allows the cells to be recovered for further studies if desired. Differential counts

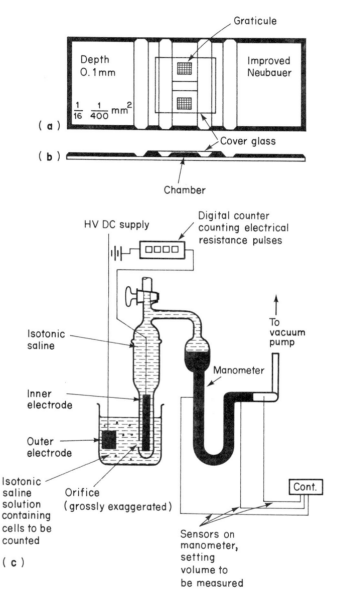

Fig. 2–2 (a) is a plan view, and (b) an elevation of a Neubauer haemocytometer. (c) is a diagrammatic representation of the essential working parts of a Coulter counter. The mercury in the manometer is drawn back over a set of electrical contacts, starting and stopping the counting process over a 0.5 or 0.1 ml volume, the cell suspension being drawn through the orifice by this action.

can be attempted, e.g. of viable cells and dead cells, whereas electronic counting will count dead and live cells alike. Of more dubious benefit is the possibility of taking this differential counting of viability one stage further by *vital staining*, using dyes such as trypan blue or nigrosine. Dead cells are thought to take up dye because of changes in membrane permeability, whereas living cells do not. Reliance should not be placed upon this assumption (see section 5.2.3).

Electronic counting is, however, both faster and more accurate, the instruments being designed to count any particle falling within a chosen size range. The counter draws a set volume of an appropriately diluted suspension of cells in isotonic saline (0.15 M NaCl) through a very fine orifice and along a tube of known dimensions (Fig. 2–2). A high potential difference is placed across the ends of the narrow tube and a DC current flows along it by means of the ions in isotonic saline. Each cell or particle passing down the tube momentarily increases the resistance and generates a signal which is amplified, analysed by a scaler device for sizing and counted. The total number of cells passing through the tube from a measured volume of suspension is corrected for dilution to give the concentration of cells per ml. The counting device is often coupled with an instrument which plots the size-distribution histogram of cells being counted. This is easily achieved because signal size is proportional to cell volume, and the instrument can be calibrated with populations of latex spheres or spores of the appropriate size range.

Remembering that the volume of a spherical object, such as a suspended cell, is $\frac{4}{3}\pi r^3$, let us suppose that a cell of 7 µm in radius (i.e. 1437 femtolitres in volume; $fl = \mu m^3$) doubles in size and divides. Immediately before division, it will have doubled its volume to 2874 fl, and its radius will be 8.8 µm. While the volume has increased by 100%, only a 26% increase in radius is seen. Although the electronic counter senses the doubling of volume with great accuracy, the human eye looking down a microscope finds considerable difficulty distinguishing cells of 7 µm radius from those of 8.8 µm radius, especially when they are mixed with cells of intervening radii. Optical measurement using calibrated micrometer eye-pieces can produce good results with rod-shaped organisms however. This is because the cells grow at the end only and tend to grow linearly from one unit length to two. A good example is the fission yeast, *Schizosaccharomyces pombe*. For accurate work on small cells of these shapes, photography at known magnifications followed by straightforward measurements of the images can be used. Computerization has also made it possible to use quick *planimetry* methods of determining cell diameters, areas, perimeters, etc., by projecting microscopic images on to a digitising tablet. A sensor probe is run around the outlines of interest, and the information is fed directly through a mini-computer, programmed to carry out appropriate calculations from the planimetry.

Counting cells on culture vessels or histological slides requires a different approach. It is necessary to use either an eye-piece graticule or some means of marking out known areas within which to count the number of cells present. For comparative work exact areas are not required as long as the same ones are used for all samples. Furthermore all the cells in a microscopic field need not be counted, as a simple *point-counting procedure* will suffice. This is a means of approximating the number of cells by the frequency with which they coincide with regularly spaced bars or randomly placed dots, since this will be proportional to the cell density. Point-counting procedures allow a large number of visual fields to be quickly scanned with a high degree of accuracy.

Quick estimations of the growth of cells in suspension can also be made by *turbidimetry*, which involves passing a beam of light through a known width of culture under standard conditions. A photocell on the other side of the vessel picks up the transmitted light, some of which will have been lost by scattering. The amount of scatter will be proportionally related to the cell concentration in the culture, and can be read against a reference set of dilutions. This technique is widely used in bacterial work and for cultures of fast-moving organisms, such as ciliated protozoa or flagellated algae.

2.2.2 Population size distribution

Since the size of a cell increases as it proceeds from one division to the next, an asynchronous population will have cells ranging from the smallest immediately after division, to those at least twice as big which are about to divide. At division, *one* cell is suddenly converted into *two* smaller progeny of half the mass and volume. Thus the sizes of cells in an asynchronous population do not exhibit a normal or Gaussian distribution. Because of the birth of two cells from every one which divides, there are more than the expected number of small cells, giving a

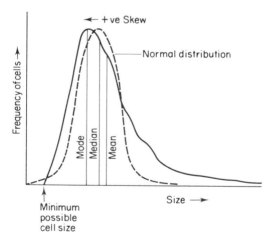

Fig. 2–3 Size distribution curve of an exponentially growing population. The normal distribution (broken line) has no skew, and mean, mode and median have the same value.

positive skewedness to the curve (Fig. 2–3). Furthermore, cells have a smallest size compatible with viability, whereas at the top end of the range a maximum size cannot be set.

2.3 Cell kinetics

Assuming the average cell size remains constant in an asynchronous culture, increase in cell number gives an accurate estimate of growth. Thus one bacterium produces two, two become four, four go to eight, and so on. The shape of the growth curve will depend on the pattern of division, of which the exponential type is one of several (Fig. 2–4, Pattern **a**). Pattern **b** in Fig. 2–4 shows linear growth, the slope of the curve depending upon the rate at which cells pass through their divisions. Linear growth is found in such places as the basal layer of the skin in the adult animal. Continual divisions of cells in this layer replace the

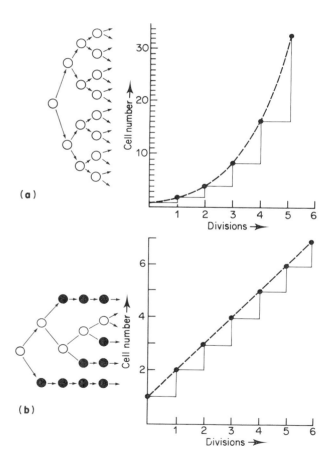

Fig. 2–4 Growth patterns. (a) Exponential; (——) synchronous, (– – –) asynchronous. (b) linear.

loss of cells which are being sloughed off the skin. When a cell in the basal layer divides, its chances are one in two *on average* of either migrating and differentiating as a keratinized skin cell or remaining on the basal layer to divide again some time later ('on average' because on some occasions both cells migrate and on others neither cell migrates, in addition to the twice as probable outcome that one stays and one goes). As long as half the cells which divide stay behind, a *stem-cell* population is present which is capable of maintaining tissue homeostasis (Fig. 7–2). If all the progeny of such a stem cell population could be counted, the curve shown in Fig. 2–4 Pattern **a** would be obtained. It is important to keep in mind that sometimes only a part of the whole population, the progeny over many generations, is actually seen. Zero growth rate of a tissue such as the skin does not mean no growth is going on, it means that the rate of production of cells is exactly matched by the rate of loss (see section 7.7). In the chosen example, skin cells will quickly change from Pattern **b** to **a** when a wound is made.

More complex curves are basically variations on the exponential and linear growth themes, due to some modifying factors which need to be discussed in more detail. In an exponential culture, the shape of the growth curve will be smooth and truly exponential as long as (*i*) *the growth fraction* (proportion of cells in population actively proliferating) is constant, and (*ii*) division occurs at a constant rate. Although the slope will alter (see Fig. 2–5), exponential curves will

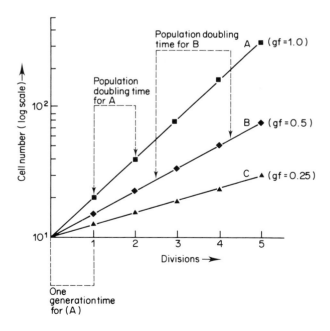

Fig. 2–5   Growth fraction (gf) and its effect on proliferation rate.

be found whatever the growth fraction is, but the equation will change from:

$$y = n^t$$

when all the cells are dividing (y being the yield, n the number of cells at the outset, t the constant time of successive divisions), to:

$$y = n^{0.5t}$$

when only a half of the cells are involved in division. A similar slope also results with a growth fraction of 1.0, but where cells taking twice as long to go through successive divisions. The interval between divisions is not invariable, even in a bacterial culture kept under optimal conditions, for reasons which will be explored in later chapers. With age, prolonged culturing, or when nutrient supplies are limiting, interdivision times become protracted. The interval between successive divisions is generally called the *generation time*. Under optimal conditions, each cell type will have a characteristic minimal generation time. In a bacterial cell culture where the generation time averages, say, 60 min, it is quite normal to find that it takes about 65 min for the concentration of cells in a culture to double. The additional time is a typical feature of the *population doubling time*, which should not therefore be confused with the generation time. The reason for the difference is that estimates of generation times are often biased in favour of the faster-dividing cells in a population, whereas cells which divide extremely slowly tend to be ignored. The greater the spread of generation times in a cell culture, the bigger the discrepancies with the population doubling time. Another reason is that the growth fraction is rarely 1.0; and a third is that not every division results in two equal and perfectly viable progeny. It is difficult to make allowances for these modifying influences as they vary from one occasion to the next. The drop-out of cells from a population or recruitment into it can dramatically alter the overall kinetic behaviour of a cell system. The impact and importance of these modulations in the behaviour of cell populations are discussed in the last chapter.

2.4   Single cell kinetics

Do cells in isolation behave the same way as those in populations? The importance of this question is that it makes us examine the intrinsic ability of a cell to grow, divide and perform other functions, free from the restraints and influences of neighbours. Various methods have been adopted to isolate and keep selected cells in view; for example, some cells have been confined in narrow capillaries, others have been plated out in small microdrops, or sometimes a meshwork of fine fibres has been used to trap cells and prevent them moving very fast. Even mammalian cells can be penned up individually when placed on little areas of wettable substrate masked onto a non-wettable substrate such as cellulose acetate.

The microculture, whether it contains a ciliate or a mammalian fibroblast, can be followed for only a short time, perhaps a few generations, because it may quickly outgrow its resources. Some indication of exponential or linear growth may be seen, but usually this procedure is adopted for more careful analysis of delays in division itself, or of behaviour of cells shortly after some treatment.

With fast-moving cells, and sometimes with slow ones, it is difficult to record the behaviour after the first two divisions because the investigator can quickly lose track of which cells are siblings, i.e. genealogy gets more complicated and difficult to follow. *Time-lapse cinematography* is then required which allows a permanent genealogical record to obtained of all activity in the culture, and analysis can be done at leisure.

The composition, size and behaviour of cells has been briefly dealt with. The next major task will be to examine in much closer detail the underlying process of how one cell grows and becomes two. You will recall that emphasis was placed on the idea that cell growth in mass is a highly ordered process (section 2.1.4); the next two chapters will be devoted to the sequence of events which occurs in *interphase* (Chapter 3) and *mitosis* (Chapter 4), which together amount to a complete *cell cycle*.

3 Co-ordination of Cell Growth and Division – The Cell Cycle

3.1 The cell cycle

Proliferating cells grow from the end of one division to the beginning of the next. There are a few exceptions in which division can occur without a cell getting bigger, such as the early divisions of sea urchin or frog eggs. Eggs may get much larger before fertilization, and can easily coast through 4 or 5 divisions without using up all their reserves. With other cells such as *Tetrahymena*, however, the daughter cells are on average half the size of the parents at division, as seen in Fig. 3–1. The fact that these cells, like most others, grow back to the parental size

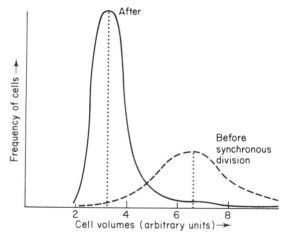

Fig. 3–1   Size distribution of synchronous *Tetrahymena* population immediately before and after division. Volume is halved and frequency doubled after division.

before undertaking a further division does not necessarily mean that there is a steady increase in all cellular components with time, nor an increase on the logarithmic pattern which a whole cell population will follow (Fig. 2–4a). Growth through the cycle is a much more carefully arranged and integrated process, as first began to be appreciated in the early 1950s when it was found that the cell made new DNA at a discrete time between divisions, called 'S' *phase* (Fig. 3–2). This was the origin of the cell cycle concept.

Before S phase begins, most cells spend some time in G_1, the G standing for gap, and not growth although increase in size occurs in most cell types. Similarly

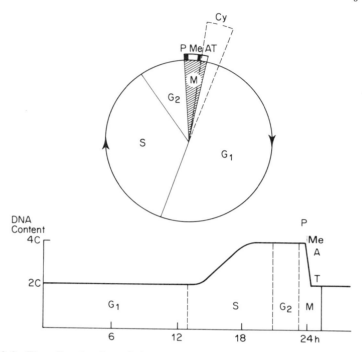

Fig. 3–2 The cell cycle of a typical mammalian cell dividing once a day. P, prophase; M, mitosis; Me, metaphase; A, anaphase; T, telophase; Cy, cytokinesis. The lower curve indicates the change in nuclear DNA content in one cycle.

after S, a second gap G_2 precedes mitosis. The period of division is either referred to as 'D' which covers the whole mitotic and cytokinetic processes, i.e. nuclear division and cytoplasmic division respectively, or simply as 'M' to indicate it is the period in which mitosis occurs. The four phases make up the familiar cell cycle, although it is not a truly cyclical event since this would suggest that each dividing cell plods on again through exactly the same routine with the same timing as in the previous cycle. It may be better to think of it as a spiral in which each turn can show considerable variability. This matches more closely the notion that it is really a continuum of activity in which the events can to some extent be shifted relative one to another. Some cannot, however; it can easily be imagined what would happen if DNA synthesis occurred at the same time as mitosis! But aside from the essential condition which prevents the genome from becoming hopelessly tangled up in replication while dividing, considerable flexibility is seen in the cycle pattern. In rapidly proliferating cells S can occupy almost the complete inter-division period. Put another way, all the other events are crammed into the same period of time that it takes to replicate the genome. This raises the small but important point that during the time DNA is being replicated, the cell carries on with its other preparations for division, i.e. S phase is not exclusively devoted to DNA synthesis. In some cases S is shifted to the very beginning of the cycle, and in others it may take place much later with division occurring almost as soon as replication has been completed. But this is not all, for in bacteria, DNA synthesis can be occurring as cells pass through division. This does not contradict the condition stated above, since DNA replication relevant to the division in progress has already been completed. In these circumstances the chromosomes have begun replicating themselves for one or even two divisions ahead by starting in the previous cycle. Chromosome replication in a bacterium may take 40 min, and yet the cell generation time can be 30 min. This is made possible by each new round of DNA synthesis starting about three-quarters of the way through the preceding cycle.

3.2   Increase in mass during the cell cycle

In order to grow, a cell must synthesize more macromolecules than it destroys. Since about three-quarters of all the macromolecules are proteins, the growth rate can be expressed in net terms as the amount of anabolism less the amount of catabolism of protein molecules occurring over a given period of time. Measurement of accumulation of biomass was described in section 2.1. It has been difficult to get accurate measurements of net gain (growth rate) in order to see the true kinetics of the process at the individual cell level. In many cells, growth between divisions is closer to a linear rate of increase than the expected exponential rate of increase (Fig. 3–3). This is one reason for the culture growth curve (Fig. 2–4) being a series of linear segments which combine to give an exponential curve. In *Amoeba*, linear growth occurs for most of the interphase period with a slowing down prior to division. The opposite is found at the end of the L-strain mouse fibroblast cycle. Some become linear, shortly after the start of growth such as *Schizosaccharomyces pombe* (Fig. 3–3) or a bacillus. Since these have a cylindrical shape and grow at one end, increase in length is a good measure of mass increase. It remains a bit of a mystery in organisms, other than those that are like *Schizosaccharomyces*, why growth does not show exponential kinetics (compound interest) and this suggests that there is some important regulatory mechanism(s) which may not only act on anabolism and catabolism independently but keeps them in the correct relationship to some other cell function(s). The major groups of biological molecules (phospholipids, carbohydrates, nucleic acids, proteins, etc.) have been studied in different cell types to see how they increase in relation to overall growth. For example, it has been shown in *Euplotes*, a ciliated protozoan, that protein increase follows the same curve as overall growth, and that the production of rRNA and mRNA would be closely related. DNA, of course, does not keep in step but shows the doubling in amount within a shorter period between divisions. Certain enzymes have been seen to change their levels through the cycle in relation to on-going functions. But for others, controversy still exists over whether these fluctuations are truly cycle-related or artificially induced by the manner in which a culture of cells is manipulated prior to the assay procedures (see section 5.4 for a fuller discussion).

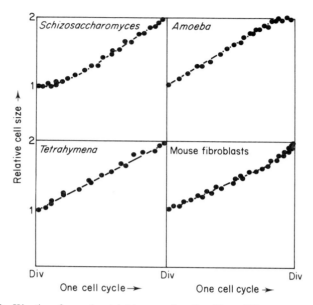

Fig. 3–3 Kinetics of mass (protein) increase in cells of four different types through one generation time.

3.3 G₁ phase

The G_1 phase begins as soon as a cell competes division except those embarking immediately on DNA synthesis. And it is a moment of decision as to whether the cell progresses through another cycle or stays in G_1 and becomes differentiated. Most of the evidence suggests that the fate of a cell is already decided somewhere during the previous cycle, probably about the time it entered DNA synthesis (S-phase). This is because it takes an S-phase plus a mitosis to 'fix' a new mode of expression in cycling cells.

For the cell which continues to cycle, the early events of G_1 include the synthesis of membrane components. There is an increase in phosphatidylinositol and phosphatidyl ethanolamine synthesis, made possible by a considerable increase in the availability of phosphatidic acid in the cell. This is particularly noticeable in cells which have been stimulated from a quiescent G_1 state back into cycle. Other studies have also suggested that the amount of phospholipids in the membrane increases steadily throughout the cell cycle with little fluctuation in levels. The problem comes with division because of the simple relationship that when a spherical object divides into two approximately equal parts, the surface area of the two smaller spheres is considerably greater than that of the original one. The factor is about 1.6. During growth, the surface area of a spherical cell would not rise as rapidly as its volume. The sudden change accompanying division (cytokinesis) must be met by either a stretching of available surface membrane, or a taking up of slack produced in interphase. The cell in interphase continues to synthesize membrane components which help to meet the demands at cytokinesis. Cells approaching division may have many processes (finger-like microvilli or spade-like projections) which can be retracted at mitosis to give surface-covering capacity. These changes are easily followed by scanning electron microscopy (Fig. 3–4). A great deal of the concern

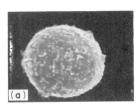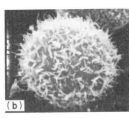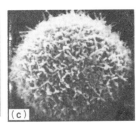

Fig. 3–4 Scanning electronmicrographs of the surface of mammalian cells at (a) early G₁; (b) S-phase; (c) late G₂. (By courtesy of the Rockefeller Press and the authors: Knutton *et al.* (1975). *J. Cell. Biol.*, **66**, 568–76.)

about the process of clothing the new G_1 cells after division is not really justified because insufficient account has been taken of the natural fluidity and elasticity of the cell membrane. Cells can be swollen and shrunk by means of osmosis without necessarily rupturing them. The amount of precursor molecules present in the cell is not known for certain but the cell membrane mass can turn over very rapidly. There may always be enough to allow insertion of new material into the membrane as and when it is required, e.g. in going from a spherical cell to a long fusiform cell. But the surface of the cell is covered with many different types of molecules, glycoproteins, sialic acids, and so on. These can also be followed through the cycle. Glycoproteins, for example, are synthesized more rapidly in G_1 but from S phase onwards the levels fluctuate. Changes of this kind are often closely followed by the number and density of antigenic determinants on the cell surface through the cell cycle.

Steady linear growth is one of the few features of most G_1 cells. The length of time a cell will spend in this phase can be highly variable. One suggestion has been that a random event occurs in G_1 before a cell can progress through the cycle. This *'transition probability' hypothesis* is a topic for discussion in Chapter 6. Accepting that a cell will pass round the cycle, events in G_1 may be very numerous and require proper sequencing or interdigitation to allow smooth progression. Until recently, no obvious 'events' had been recognized, but the use of temperature-sensitive mutants, cells which have been mutated with chemicals such as MNNG (N-methyl-N′-nitrosoguanidine) and then selected for whether they will grow or not grow at certain temperatures, has made it possible to detect certain end-points which seems to be essential for a cell to maintain its impetus in the cycle. Mutants can pass these 'end-points' at normal temperatures but become stuck at non-permissive temperatures (see section 5.8). In the same manner as in classical genetics, the unfolding of these processes can be mapped as G_1 events of the mammalian cell cycle, much as they have been mapped on a bacterial chromosome.

One point used in these analyses is the time at which DNA synthesis (S phase) is initiated. By use of mutants or chemical inhibitors known to interfere specifically with metabolic processes, cells can be treated to see if they are competent to start DNA synthesis. Prior to the onset of replication, it would seem logical to us (although the cell has no such 'logic') that the enzymes required for this process should be available. There are certain enzymes, such as thymidine kinase, thymidylate synthetase and others, which increase in cells just before DNA synthesis begins. Adding inhibitors of protein synthesis will stop this occurring and arrest cells at the last part of G_1 or as soon as they begin to enter S-phase. A cell which has the basic requirements for initiating DNA synthesis has passed the point of *commitment*, but what the signal is which triggers progression into the next phase remains obscure. Before the onset of S phase mRNA synthesis is needed, as can be shown by RNA inhibitors stopping cells going into S, and similarly for protein synthesis. Because of the dependence of the latter on the former, the *transition point* for RNA synthesis must be before that for protein synthesis. However, these analyses have the drawback that they may seem to act on a simple end-point, whereas many minor ones may be involved on the way. They do not tell one a great deal about the places in the preceding phase at which they may be arrested. The transition points mentioned above should not be considered as milestones around the cell cycle. This is because their position can be shifted quite considerably by the conditions under which the cells are growing. The dose levels of inhibitors used in such analyses can by themselves cause considerable differences in the determination of the transition point, the higher the dose, the earlier in the cycle the action on cell cycling occurs. This is further complicated by each inhibitor producing its own specific effect. For example, one can apply several different inhibitors of protein synthesis at equal potency (% inhibition) and get several distinct time points for the transition of cells from G_1 into S phase. In the world of yeast and bacteria, the expression transition point is known as the *execution point*. The event which inhibitors can prevent a cell reaching, or the last action it is capable of undertaking after treatment, is called the *termination point*. At present, the G_1 phase of the cell cycle can be one of the longest and yet the least understood in terms of distinctive events required for the movement of cells through their cycle.

3.4   S phase

There is a lot of DNA in a bacterium, about 1000 μm in the form of a circle. The mammalian cell contains a great deal more, and in each case it must all be replicated in an orderly fashion during S phase. In the bacterium, there is a point on the bacterial chromosome, some triplet sequence on the circle of DNA, at which replication starts and then proceeds along each of the two separated strands of DNA in opposite directions simultaneously. In the mammalian cell, there are probably as many as 2000 initiation sites, not all of which are set off at the start of S phase. The pattern of initiation in the more complex genomes usually follows the same pattern at each cycle. Semi-conservative replication is involved requiring the separation of the two strands of the parental DNA of the Watson–Crick double helix, which exposes two anti-parallel strands ($3'$–$5'$/$5'$–$3'$) on which the complementary bases (nucleotides) can be polymerized by a complex set of enzymes which can be briefly referred to as replicases. There are a whole host of other enzymes in attendance to supply the appropriate deoxyribonucleotides, to open up the DNA, to remove associated histone or non-histone proteins, to nick the strands, etc. The complexity of the process makes it obvious that some of the latter part of G_1 will have been concerned with some of the preparations for this event. In the mammalian, as in the bacterial chromosome, there are short segments made on the strand running $3'$–$5'$, and slightly longer segments on the $5'$–$3'$ strand. These are subsequently joined together by enzymes called ligases. But the whole circular chromosome is formed in this way in bacteria, whereas in the mammalian chromosome, there may be a part of the tortuous DNA molecule made at one time, and much later another section of the same chromosome may be replicated. Interestingly, one can follow this by studying the labelling pattern of chromosomes with a specific DNA precursor such as tritiated thymidine in each cell cycle. A similar pattern of replication is picked up for each part of the chromosome. Certain parts are always replicated in early S, others later in the S phase. Some chromosomes are always very late in their replication, but exactly how this pattern of synthesis is controlled is not known. It is repeated with remarkable consistency cycle after cycle.

It is wrong to think that the S-phase cell only replicates DNA. There is continued growth, changes in the surface morphology, and multiplication of subcellular organelles. As the DNA is made, a requirement arises for proteins which are characteristically associated with it, notably histones and also some non-histone nuclear proteins. The former are highly basic proteins containing a preponderance of arginine and lysine (dibasic) amino acids. One histone (H1) is lysine rich and is concerned with the spacing of DNA in bead-like structures (nucleosomes). The other histones (H2–H5) are concerned with the covering of DNA and supposedly with its availability for transcription, i.e. the reading of the code to form the corresponding mRNA. Obviously DNA which is being replicated cannot be transcribed, but it is interesting to consider that the DNA structure must be opened up in order to be accessible to these processes and that at one time the DNA replicases must get in while at other times the RNA transcription enzymes must have access.

There is a very pronounced acceleration of histone synthesis during S phase, while at other times in the cell cycle it is very much slower. These proteins are made on the ribosomes in the same way as all proteins, and not in the nucleus as originally suggested. They must move from the ribosomes to the nucleus to be titrated against the newly synthesized DNA, but how this translocation occurs is not known. It is particularly noteworthy that arresting DNA synthesis stops histone synthesis much more quickly and completely than stopping RNA synthesis. Normally switching off RNA synthesis stops messenger being produced thereby halting protein synthesis, but the run down takes some time. The evidence suggests that there is a sensitive feed-back mechanism from site availability in the nucleus. When DNA replication is stopped, no denuded DNA is left and the resulting excess of histone rapidly turns off its own synthesis.

Once DNA synthesis has started in a cell which has an adequate supply of nutrient, it continues until it is complete, although the rate at which it proceeds at first accelerates, reaches a maximum and then gets slower towards the end of S phase. The heterochromatin in the nucleus loosens up to give the appearance of euchromatin, and the replication may occur in the peripheral region of the nucleus, possibly close to the nuclear membrane. Since there is a set amount of DNA in each diploid nucleus at the outset, the time it will take to complete S phase can be calculated if the average rate of replication is known. It does, in fact, turn out to be a fairly constant time in each cell type under normal conditions. Since the pattern of replication is also predictable, as described above, it follows that S phase is a tightly-controlled period within the cell cycle. For this reason it is of particular significance for the cell kineticist measuring the proliferative activity of cell populations (see section 5.3).

When DNA synthesis has been completed and a 4C complement of DNA installed in the nucleus, the cell can proceed to division. In some cells this is done without delay, whereas in others this is not enough. We have already mentioned that there is some kind of suppression of division when the cell is replicating its DNA although the cause of it is unknown. It is difficult to force an S-phase cell into mitosis, but certain fusion techniques have now become available (see section 5.7), which may suggest that there are other influences holding back the cell from entering mitosis.

3.5   G$_2$ phase and the initiation of mitosis

The period from the end of S phase to division is poorly understood, although it can be clearly delineated. The technique of analysing it is either to label cycling cells and study those arriving in division with the radioactive tracers for DNA synthesis, or simply to add inhibitors of DNA, RNA or protein synthesis and follow the ability of cells to reach mitosis. In the latter case, cells cease to move out of S phase after addition of the inhibitor and therefore a clean cohort of G$_2$ cells will pass on into mitosis. By erecting other blocks at intervals along this time, a crude analysis of events in G$_2$ can be made. It is crude because there are very few compounds which can be specifically applied to cells in G$_2$ which can tell us something about what is going on. The data suggest that for a period after the start of G$_2$, both protein and RNA synthesis are needed. Shortly afterwards RNA synthesis is no longer required and, closer to mitosis itself, protein synthesis is no longer necessary, i.e. the cell has reached the transition point in each of these cases where it can coast into division on what it has already made. Use of different inhibitors claimed to have the same effect on say protein synthesis can often give quite different transition points and therefore the data must be viewed with caution. Here as in all other cycle analyses, the problem of deciding the end-point to be measured has to be carefully considered. By far the easiest way of analysing G_2 events is to add inhibitors to cell growing cultures or tissues and see when they become fully effective in stopping cells increase in number. It has already been stressed that some cells do not actually separate until some time after they have entered the next interphase. Therefore end-points relying on the cessation in cell number increase are not going to give a very accurate transition point for a drug thought to act in G_2; this result could be true of inhibition at any time during the whole $G_2 + M +$ cytokinetic period. In more reliable studies, the ability of cells to enter mitosis, specifically prophase or metaphase, is measured (Fig. 3–5).

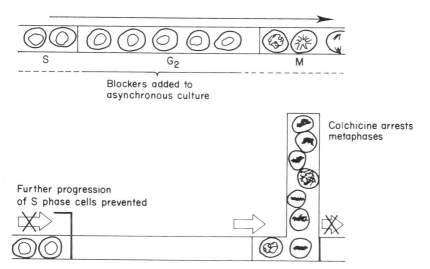

Fig. 3–5   Analysis of G_2 progression. Adding the blocker at a given moment (top) will stop S-phase cells progressing but G_2 cells pile up in mitosis (bottom) due to colchicine.

The cell that is fully prepared will enter division, but it is not known what makes it finally ready. Is it its absolute size or the right concentration of an important molecule? There are many other equally plausible possibilities. One is that at the end of DNA synthesis, the cell actually begins to enter mitosis. Since little attention is paid to the subtle changes of the nucleus during this period, the cell is still considered in an interphase condition. A second possibility is that although the DNA is replicated by the end of S-phase, it has to be edited to remove the mistakes made in copying, otherwise an unacceptable level of genetic aberration would be passed on to the daughter cells. When this process is complete, the cell proceeds immediately into mitosis. Perhaps the best known hypothesis is that a specific trigger mechanism exists which sends a cell into mitosis. This could be influenced by signals sent from other cells or from within the cell itself. But we are left asking the question what pulls the trigger at the right time. Current opinion would suggest that no single event controls the passage of a cell from G_2 into mitosis. It is probably caused by the concurrence of a number of discrete processes meshing together when the condition of the cell has reached a state of preparedness to enter division (see Chapter 6).

Since there has been considerable concern about the synthesis of molecules, the DNA, proteins, membrane components, etc., required for division, it should be remembered that a cell is not just a bag of macromolecules. It has a highly organized structure arising from the regulated interactions of RNA, protein, lipids, polysaccharides and all the other molecules. Their assembly into structure is as important as their synthesis. During G_2 it is known that synthesis alone is not enough to initiate division. Assemblies are made and these seem to be particularly sensitive during their building phases. It would appear that events of this nature are critical in G_2 (section 6.3).

Many G_2 animal cells round up as they go into division. Some do not, notably marsupial cells. In those changing shape, the fine filamentous and microtubular structure of the cytoskeleton of the cells retracts and the molecules are redeployed for uses in mitosis. As this begins, re-assembly of these structures is occurring in other parts of the cell. This is a good example of diametrically opposed reactions occurring in different parts of the cell. The cell proceeds with a most remarkable revolution which virtually turns everything topsy-turvy, yet, by a very subtle control of these opposing reactions, order is restored and two new interphase cells looking very like the parent are formed.

Cells will not proceed into division if they have been irradiated beforehand. Exactly how much of an interval between treatment and arrest occurs depends on the cell type and the dose given. Some grasshopper neuroblast cells which have been studied are so sensitive that as little as 4 rads of X-irradiation will send them back from the brink of mitosis. The reason that irradiation is mentioned here is that it is one of the few agents which can arrest cell development very late in the cell cycle, and so far there is no clear explanation of how this happens. It could be that it makes another break or aberration in the chromosomal material which has to be repaired. This idea is not particularly good because cells can be treated with quite devastating levels of other DNA damaging agents at these late stages of G_2 and yet not impede their progress into division. From fundamental considerations concerning the number of X-rays needed to cause arrest – the 'single-hit' nature of the phenomenon – and the fact that it is the only organelle within the cell which could have the right geometry, it is now suspected that the target organ is the nuclear membrane, the site of chromosome condensation (see section 4.1).

4 Mitosis and Cytokinesis

4.1 Prophase

When a cell enters division it undertakes an extensive rearrangement of its cytoplasm and nucleus so that the chromosomes can be segregated between the two new daughter cells. Although the mitosis can sometimes appear to be precipitated by various external influences, a cell cannot normally enter division unless it is adequately prepared. In many cases, an external stimulus responsible for initiating division cannot be found, and this implies that the driving force comes from within the cell itself. External factors, on the other hand can inhibit progress, as seen with irradiation. Although it would be convenient to think that there is a specific factor or reaction which initiates mitosis, it is probably more accurate to imagine the process as being similar to the opening of a combination lock with several steps having to be undertaken in strict order or simultaneously to set things going (sections 6.3 and 6.4).

Most cells which have progressed into G_2 of the cell cycle reach division within a fairly well circumscribed period of time. Few can reach it quicker than a certain minimum time, suggesting that there are definite preparations to be completed in this phase. Cells in certain tissues (e.g. kidney) seem to hold up in G_2 for relatively long periods of time. A proliferative stimulus to such tissues, e.g. the removal of one kidney resulting in hyperplasia of the contralateral kidney, brings these cells into cycle again. They enter division very quickly and act as a first line defence to combat the sudden deficiency status. They must still complete their G_2 functions but there is one documented case of cells being stimulated to enter division faster than they would reach it in normal cycling. This was claimed for nuclear divisions in the syncytial plasmodium of *Physarum polycephalum*, a slime-mould. A phosphokinase preparation taken from tumour cells was placed on the fungal plasmodium, causing the nucleus to enter division within the equivalent of only 70% of the normal cycle time.

In cells of growing populations such as the basal layer of the skin or ascitic tumour cells floating freely in the peritoneal cavity, it is interesting to find that although their rate of proliferation is very much faster than that of liver cells and perhaps only a little slower than the crypt cells which make the lining of the gut, mitosis itself occupies a similar length of time in each. For most mammalian cells this is about 30 minutes to 1 hour. But there is considerably more variation amongst cells of different species, both in the time that division takes and the manner in which it occurs.

In prokaryotes, such as bacteria, the nucleoid lengthens, constricts in the middle and the two new nucleoids form. This is followed by septation between them, i.e. a wall is formed in the mid region of the cell, and the two new offspring can move apart. Sometimes they do not separate and end up as strings of cells.

Eukaryotic cells such as yeast undergo an intranuclear division of their genetic material, with much less involvement of the cytoplasm until separation occurs. In higher eukaryotic cells, the cytoplasm is more heavily involved and a great deal more reorganization occurs.

One difficulty with prophase is knowing when it has started. Subtle changes in the *chromatin* of the nucleus (the nucleic acid and its associated proteins) occur during S and G_2 phase, and these may be related to the setting off of condensation of the chromosomes. Some careful investigations have described in great detail the fine changes in the nucleus and cytoplasm throughout the whole cell cycle. It might be possible to pin-point the stage reached by a cell in this way with much more precision than has hitherto been suspected. But these changes are a reflection of metabolic processes going on inside the cell, and the condensation of the chromosomes is probably due to the protein molecules associated with the nucleic acid becoming modified. Just before condensation begins, the attachment of side-groupings – by methylation, acetylation and phosphorylation – is found to fluctuate considerably. The phosphorylation (or possibly a sudden dephosphorylation after a phosphorylation) has been considered instrumental in setting off chromosome condensation (Fig. 4–1).

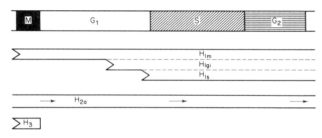

Fig. 4–1 Histones start to become phosphorylated at different times in the cell cycle ($>$) and remain phosphorylated as shown. Compare histone H_{2a} (always phosphorylated) with histone H_3.

This is the basis on which the phosphokinase story mentioned above in relation to the slime mould *Physarum polycephalum* is based. H_1 histone is known to be heavily phosphorylated before prophase begins and almost immediately afterwards become dephosphorylated, although it is difficult to know whether this is a cause or an effect of the onset of division.

Chromosomes condense at prophase on to the nuclear membrane. From the evidence presented in the previous chapter, it is obvious that this organelle plays an important, but poorly understood, role in division processes. The coiling of chromosomes makes them shorten as prophase proceeds and their modified chromatin makes them become tightly packed. This accounts for their more intense staining at mitosis, whereas in interphase they are extended and poorly stained. When followed by high power light microscopy and/or carefully correlated electron microscopy, the chromosomes are seen to condense directly on to the inner nuclear membrane in definite positions relative to one another. It has been proposed that the pore-like structures seen on the nuclear membrane (Fig. 4–2) are like poppers which serve as attachment points for specific chromosomes and make an orderly hollow ball of chromosomes come to rest on the nuclear membrane. This may appear surprising because one of the next events is the disappearance of the nuclear membrane, leaving the chromosomes free in the cell. The nucleolus also begins to dissipate. The material making up the nuclear membrane and the nucleolus is not lost, however, for about three-quarters of it is re-assembled after division, illustrating a remarkably high degree of conservation of materials. A considerable amount of it remains associated with the chromosomes during division. It has been noted above that in many organisms the nuclear membrane remains intact during division, the whole process of chromosome segregation being performed within an intact nucleus. Whether the nuclear membrane plays such an important role in these cases is not known.

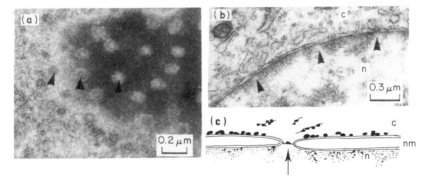

Fig. 4–2　Nuclear pores (▲) as seen in a grazing section of the nuclear envelope (a), and in sagittal section (b), electron micrographs of a fibroblast. (c) gives a diagrammatic impression of a pore with nm representing the inner and outer membranes, c, cytoplasm and n nucleus. The pore sometimes appears plugged (⚲), but very large macromolecules can easily move in and out of the nucleus.

As the nuclear membrane itself begins to fragment, the periphery of the nucleus appears to be surrounded by a zone of relatively clear cytoplasm. Within this zone there are two *mitotic centres*. Previously there had been only one in evidence, but at some point this region (the cell-centre or centrosome) has split into two. The mitotic centres become recognizable by their interaction with each other, contributing to the impression that they are forcing each other apart (see section 6.2) The two migrate to opposite ends of the nuclear zone, and each helps to make the half-spindles which interact to form the complete spindle structure. There are many problems associated with these events, such as the fundamental issue of polarity and its generation, the behaviour of the centrioles and the force moving the mitotic centres (centrosomes) apart, which show that much of mitotic research is still speculative (Chapter 6).

4.2 Metaphase

Clear borders between the phases of mitosis are not always apparent because the cell shows some flexibility in the order of certain minor events. The delineations are for the convenience of the scientist wishing to communicate about them, not to assist the cell in getting things the right way round. The early part of metaphase, *prometaphase*, is a period during which the nuclear envelope can no longer be found, the mitotic centres migrate to their final polar positions, and the chromosomes disperse within a clearish area of cytoplasm in the centre of the cell. Metaphase is reached when the chromosomes begin to organize themselves at the equator of the spindle. Their arrangement gives this configuration, the *metaphase plate*, and homologous chromosomes (like-pairs) spend some time becoming paired up. More chromosomes tend to lie at the circumference of the plate, fewer in the central region, especially when chromosome number is low. How this sorting out process occurs is unknown, but their alignment at the equator is believed to be a natural consequence of chromosomes moving into areas of least (or equal) tension within the spindle confines. Spindles are rather elaborate structures of which the basic con-structional element is the *microtubule*, the same structure found throughout the cytoplasm in interphase which also occurs in cilia and other organelles. In early studies, the elements were called fibres, and they are often referred to in this way today. When spindle formation is prevented by alkaloids, such as colchicine or vinblastine, the chromosomes remain in an unsorted bundle in the middle of the cell. In cells which are left unrescued from drug treatment, the chromosomes end up as a pycnotic mass in the centre.

Homologous chromosomes in a fully formed spindle link up with the microtubules of the half-spindles. The connections between the chromosomes and the poles are made by microtubules running from the *kinetochores* (Fig. 4–3) to the mitotic centres. There are some fibres running out from the poles which do not meet up with kinetochores, and others that have been developing throughout, going right across the central area of the metaphase plate. The latter are the pole-to-pole or continuous fibres. The formation of this elaborate structure takes place with increasing rapidity as the mitotic centres reach the pole positions although parts of the structures would have been forming from the time they started to migrate into position. For several reasons, metaphase is one of the most useful stages at which to examine cells. The cells are biggest at this time, the chromosomes are most densely staining and are highly condensed into comparatively short or stubby forms, which makes for ease in *karyotyping* and banding (Fig. 4–4). The mitotic processes can also be arrested with considerable ease at this point by destroying the spindle, as mentioned above. Metaphase lasts much longer than other mitotic phases and can therefore be a useful point at which to obtain an assessment of mitotic activity in a cell population. Also the drugs mentioned above which stop spindles forming do not

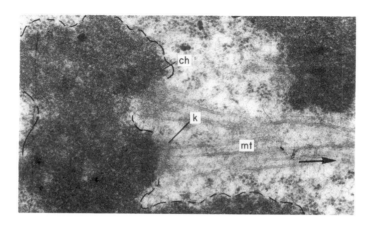

Fig. 4–3 A kinetochore (k) attaching microtubules (mt) of the half-spindle to a chromosome (ch). The arrow points to the pole towards which the chromosome would be travelling.

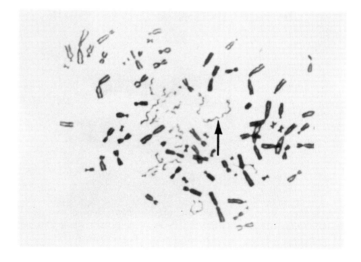

Fig. 4–4 Human chromosome preparation allowing karyotyping. It shows short dense bivalents of the mitotic human (HeLa) cell which was fused with a G_1 cell of Chinese hamster origin having much fewer chromosomes (arrow). The monovalent nature of the latter is evident (chromosomes have yet to replicate in S) by their long, extended forms. (By permission of the Wistar Institute Press and the authors: Hittelman and Rao (1978). *J. Cell. Physiol.*, **95**, 333.)

stop cells moving into division, thereby allowing a cumulative index (rate of movement of cells into mitosis) to be determined.

Before chromosomes begin to segregate, cells hold their metaphase plate configuration for some time. Although minor adjustments may be going on in the association of microtubules with kinetochores, etc., it is obvious that movement is restrained until all the chromosomes are ready, or a synchronizing signal is given to move off. Even if all the chromosomes do not completely separate at one go to opposite poles, their first intentions of doing so are clearly seen. Whether there is an active repulsion of the homologous chromosomes, or there is a shortening of the microtubules between the kinetochores and poles, the effect is for a sudden and dramatic change when the chromosomes move – almost spring – apart. The spindle at this time increases its overall length by an appreciable amount. Some of the microtubules running through the spindle can be seen to overlap with others from the opposite pole. The degree of overlap suddenly diminishes at this time (Fig. 4–5). One theory which will be dealt with later concerns the movement of the tubules against one another, rather like small muscles – the sliding filament theory (Figs 4–5 and 4–6). As soon as the chromosomes start to move apart, anaphase is said to have begun.

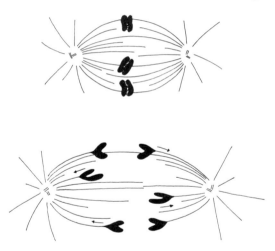

Fig. 4–5 Schematic diagram of microtubules/spindle fibres during the onset of anaphase. Kinetochore-pole tubules disassemble at the poles causing chromosomes to move as shown.

4.3   Anaphase

Unlike metaphase, there is no waiting about in anaphase. Once it has started it goes smartly ahead to a conclusion, seemingly using little energy in sending the heavy chromosomes to the opposite poles. The kinetochores of homologous chromosomes being on opposite sides of the two structures when they paired up,

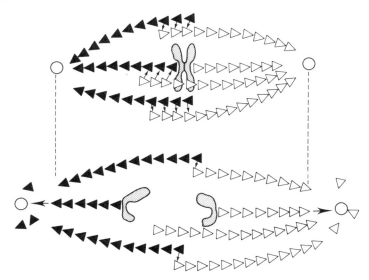

Fig. 4-6 A model of the anti-parallel 'sliding filament' idea coupled with the other mechanism of microtubule disassembly at the poles (O) during anaphase. Tubules may also increase at their ends away from the pole to lengthen the spindle.

it seems logical that the two chromosomes will attach to opposite poles and therefore be moved in opposite directions. At the start of anaphase, both chromosomes sometimes set off in the same direction only for one to be drawn back and go to its designated pole. In more artificial circumstances, chromosomes can be moved out of place within the cell by microsurgery. Some particular interesting observations by Dr Bruce NICKLAS *et al*. (1979) in this respect have shown that some link must persist between the chromosomes which have been moved out of place and their original pole. Even quite drastic relocation of a chromosome shows that it will either migrate to its appropriate pole and not just a nearer one; or if it does not make the whole journey, at least it makes movements towards the 'correct' pole. Another finding of these experiments was that the amount of energy required to move a chromosome was extremely small, making more plausible the second hypothesis that chromosome movements are due to disassembly of microtubules at the polar end (i.e. at the mitotic centres) and not to active repulsion by the chromosomes.

Occasionally chromosomes experience difficulty in separating, and they can sometimes be seen in contact at the end of one of the arms while the rest of the structures are moving poleward. This becomes a race against time because constriction of the cytoplasm between the separated chromosome sets, i.e. cytokinesis, has usually begun. Lagging chromosomes can become trapped within the constriction forming between the cells, as is seen in cases of chromosomal disturbance following irradiation, alkylating agents and other agents damaging DNA. Usually it is resolved by extra squeezing or pushing in one direction or the other, sometimes with the correct chromosome distribution resulting, at other times with the wrong arrangement.

Anaphase ends when all the chromosome arms have been folded into the new reforming nucleus close to the mitotic centres. The mechanism by which chromosomes are pulled or pushed to the poles, has taxed many of the finest brains in cell biology and some hypotheses are discussed in Chapter 6.

4.4 Telophase and cytokinesis

As soon as some gap is visible between the sets of chromosomes in anaphase, a ring corresponding to the region of the cell periphery around the equatorial plate, begins contracting. Beneath the surface of the cell membrane, a thickened area of filamentous cytoplasm is found containing fibronektins and contractile filaments such as actin. The parting of the chromosomes results in a weakened region of cytoplasm in the inter-zone through which this contractile ring creates a waist. This continues until only the pole-to-pole fibres are trapped by the contraction and two new daughters are nearly formed (Fig. 4–7). The process of

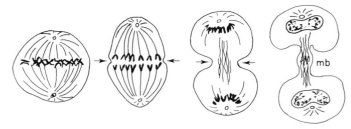

Fig. 4–7 Sequence from metaphase through to telophase showing chromosome movement, spindle changes and formation of mid-body (mb).

cytokinesis ensures that the parent cell divides into two. It is not always very accurate and some variation in the behaviour of daughters might be attributed to differences arising at division. But the important fact is that the initial gap between the chromosome sets decides where the cytokinetic furrow will form. On certain occasions, cells go through grossly unequal divisions. Where the spindle is highly eccentric, the cytokinetic furrow still cleaves the cell across the gap between the chromosome sets. This produces grossly uneven portions with one large and one small cell resulting from the division.

Careful analysis by both light and electron microscopy of the division furrow has shown that the isthmus left between the cells is like a rigid bridge. The trapped pole-to-pole microtubules of the spindle cumulate a dense material upon them (Fig. 4–8). Separation of the cells has been attributed to waves of cytoplasm passing across the *mid-body* from one side to the other, eventually causing the bridge between them to break. Other mechanisms, however, may be involved. The bridge is the site upon which reaction to the forces of separation can occur. It has also been suggested that, at the end of cytokinesis, cells must be

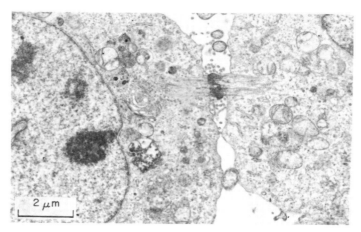

Fig. 4–8 Mid-body between two mammalian cells. A dense dilatation can be seen in the mid-region.

held firmly apart in order that they may separate cleanly. Cell membranes may otherwise fuse again in late cytokinesis, resulting in the formation of a binucleate cell the size of the original mitotic cell. This may seem an improbable event, but it might explain the increase in ploidy found in such tissues as liver in the developing mammal.

Separation at division depends to a considerable extent upon whether the cells are mobile or not. As soon as cytokinesis is in progress, some cells begin to spread out and daughters move apart. Other cells may stay as a pair until the cytoplasmic waves described above force them to separate. Huge eggs like those of the frog or sea urchin, find no difficulty in cleaving but of course they do not need to separate much in space. The above discussion shows that two quite

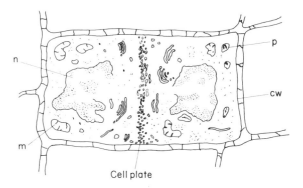

Fig. 4–9 Plant cell in division, nuclei reforming and cell plate developing. m, mitochondrion; n, nucleus; p, plastid; cw, cell wall.

distinct operations must be involved, *cleavage* and *separation*. But plant cells cannot usually move apart. They divide by the formation of a *cell plate* (Fig. 4–9) which is like the process of septation in bacteria. Clusters of small vesicles containing membrane components, which may form within the Golgi apparatus, align on the equational plate after the chromosomes have separated. They fuse until two separate membranes form between the divided cells. Later cellulose walls may be laid down between the membranes.

Telophase is, in many respects, the mirror image of prophase in being characterized by the decondensation of the chromosomes, the reappearance of the nuclear membrane and nucleolus, the dispersion of the spindle, with the mid-body elements being the last to go. Cells can take a considerable time to detach themselves, during which they move on to their G_1 functions.

4.5   Metabolism during mitosis

The cell actually spends most of its energy preparing for division, which, once initiated, proceeds to completion almost effortlessly. When production is interfered with beforehand, the deficit is either made up, or, if the cell does go into mitosis, the best compromise is made. But there is one big difference, the mitotic cell is not carrying out 'business as usual'. Considering the upheavals which occur, this is not unexpected. DNA synthesis is absent, RNA synthesis is barely detectable, and protein synthesis is usually down to about 20–30% of normal. Many cells also shut down motile systems including ciliary action.

Intracellular ion shifts may occur, the metaphase cell having a much more even Na^+ distribution. Many membrane properties alter affecting membrane potential and permeability. One ion, Ca^{2+}, and its regulator protein '*calmodulin*' are currently of interest. The ion is difficult to measure in its free form in physiological conditions but calmodulin has been followed with fluorescent-labelled antibody. Changes in free Ca^{2+}, by inference, have been seen, which may alter the assembly and disassembly of tubulin. In Chapter 6, a model based on waves of release and sequestration of calcium is described.

5 Experimental Approaches to the Study of Growth and Cycle Analysis

5.1 Introduction

There is inadequate space to give more than very brief outlines of some of the experimental methods used in studying cells and the cell cycle (but refer to AHERNE *et al*, 1977; CARTER, 1970; LUYKX, 1970; MITCHISON, 1971; PADILLA *et al.*, 1974 and PRESCOTT, 1976 in the reading list).

5.2 Cell culture

A wide range of animal and plant cell types can now be successfully cultured. The ideal situation for experimentation is to have cells growing in a defined medium. While this is easy to achieve with autotrophs (section 1.4), heterotrophs require serum, embryo extract or some other ill-defined and variable supplement.

5.2.1 Primary cultures and established cell lines

Tissues are usually diced into small (about 1 mm³) pieces and left under medium as *explant cultures*, or trypsinized to liberate single cells. These are dispersed in fresh medium and cultured at the right temperature, usually at pH 7.2. In mammalian cell *primary cultures*, two cell types emerge, fusiform fibroblasts and pavement-like epithelial cells (Fig. 5–1). Fast growing cells need

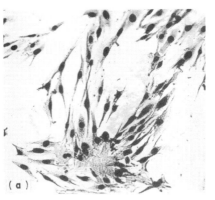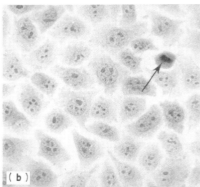

Fig. 5–1 Fibroblast (**a**) and epithelial (**b**) cell cultures. Mitosis arrowed.

early subculturing; the cells are released from the substratum with trypsin and diluted in fresh medium.

Some cells are easily propagated, others grow very little. Mouse cells often grow rapidly and indefinitely to give *established cell lines*, whereas chick cells rarely do this. Strictly, a cell line is the progeny of a single cell. Some of these have become 'immortal', such as the L-strain mouse fibroblast or the notorious HeLa cell (started from a cervical cancer 40 years ago).

5.2.2   Cell cloning

To minimize variation, cells can be *cloned*. This requires considerable skill in dispersing cells, letting individuals grow up and choosing those with the right characteristics for further propagation (Fig. 5-2). One difficulty is that isolation

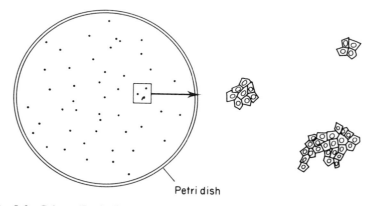

Petri dish

Fig. 5-2   Culture (Petri) dish with clones, left. Individual colonies or clones on right; under the microscope, the number of cells can be determined in each.

from one another tends to inhibit cell growth. Growth is encouraged with foetal serum, growth factors, embryo extract, conditioned medium or even a layer of lethally-irradiated cells (a feeder layer) beneath the cells.

It is also possible to separate cells with high growth potential by cloning. A neat technique of separating transformed (cancerous) cells from normal cells is to put them in 0.3% agar in medium (sloppy agar) over a firm layer of 0.5% agar in medium. Colonies of transformed cells develop in the upper layer but normal cells do not grow. This is a method which has been used to test mutagens and carcinogens.

5.2.3   Plating efficiency

Cloning is an invaluable assay of cell viability, but for a normal cell population only 50–60% of cells form colonies, not 100%. This *plating efficiency* should theoretically be equivalent to the growth fraction, but cloning conditions considerably reduce it. The technique is useful in properly controlled comparisons, and estimates the loss of reproductive viability of cells treated with toxic agents. It also allows the number of individual cells in each colony to be counted, giving a crude estimate of generation time.

5.3   Radioisotopic labelling

Aherne *et al.* (1977) give a concise treatment of this subject. In outline, either [^{3}H]- or [^{14}C]-thymidine is usually used because it goes exclusively into DNA of S-phase cells. The ratio of labelled nuclei to unlabelled can be assessed at any time by autoradiography (section 2.1.3 and Fig. 2–1), giving the *labelling index*. As the radioactive precursor treatment progresses, more cells become labelled, measuring the rate of entry into S. Eventually all nuclei would become labelled, but in practice it may be found that only 70% of the cells are cycling, thus the labelling index reaches a plateau at 0.7.

Labelling indices give only a rough guide to population kinetics. More data are obtained by a short pulse with labelled thymidine. Subsequently, autoradiographs are inspected for *labelled mitoses*, not nuclei. At first none are seen (Fig. 5–3) until an interval of a G_2 phase has elapsed. From the various limbs of the

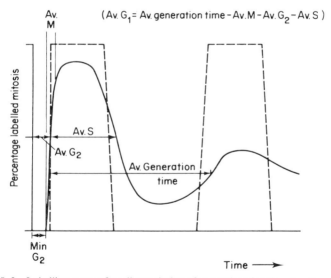

Fig. 5–3   Labelling curve of a cell population after a pulse of radioactive thymidine. The dashed line gives the shape for an idealized population of cells.

curve, average durations of cell cycle phases are deduced. Because cells show variability from one cycle to the next, the second wave is considerably dampened out (Fig. 5–3), but from the midpoint of its rise back to the same point in the first gives an estimate of the generation time of cycling cells (non-cycling cells remain cryptic).

5.4 Anabolism and catabolism

DNA accumulation measured with labelled precursors during the cell cycle (Fig. 3–2) shows a rise from G_1/S and a slowing down to negligible levels by the end of S. Histone synthesis follows the same pattern. But for most *constitutive* molecules of the cell, there is usually a more constant rate of production and a steady accumulation through the cycle. Growing cells synthesize a surplus of proteins and the sorting of this in cell construction leads to appreciable breakdown or catabolism.

Some macromolecules, such as thymidine kinase, appear for only a short time in the cycle, and others are *inducible* rather than constitutive. As they are enzymes, biochemical analysis is carried out, but this requires plenty of material usually from synchronized cell populations. Four main patterns are seen:

(i) Enzymes which keep pace with the increase in cell mass as it grows, showing a linear or a smooth increase in total activity through the cell cycle.

(ii) Those which are similar to (i) but differ in showing a definite change of rate at some point in the cycle, usually at the time S phase is occurring. One explanation is that the number of genes in the cell doubles as a result of DNA replication. Therefore potentially twice as much mRNA can be transcribed and twice as much enzyme formed (the gene-dosage hypothesis).

(iii) Enzymes which suddenly rise in activity at some point and sustain the new level thereafter – step enzymes.

(iv) Enzymes which show a sharp rise in activity at some point in the cell cycle but which equally quickly show a decline to the original level – peak enzymes.

When the pattern of proteins synthesized by cycling cells is carefully inspected by electrophoresis during the cell cycle, remarkably little change can be detected. Although there are undoubtedly massive changes in the activity of some type (iii) and type (iv) enzymes, there has been some objection that synchronising procedures used in many of these experiments are responsible for enzyme fluctuations.

5.5 Inhibitors

Both physical and chemical inhibitors can be useful in cell cycle analysis provided they are specific in action and do not produce disturbing side effects (Table 1). Few meet this ideal, but as long as their limitations are borne in mind, they continue to be immensely valuable.

Some inhibitors act immediately, others require time to reach an effective concentration. Cells which are beyond a specific point in the cycle (transition point) are not stopped whereas those approaching the inhibitor's arresting place

Table 1 Inhibitors of the cell cycle.

Main target	Inhibitor	Mode of action
DNA synthesis	Methotrexate	Anti-folate inhibitor
	Hydroxyurea	Inhibits ribonucleotide reductase
	Excess thymidine	Feedback inhibition
RNA synthesis	Actinomycin D	Blocks mRNA transcription
Protein synthesis	Puromycin	Causes chain termination
	Cycloheximide	Inhibits chain initiation
	Histidinol	Blocks tRNA^{His}
Antitubulins	Vinblastine	Inhibits microtubule assembly
	Colchicine/Colcemid	"
	N$_2$O under pressure	"
Cell separation	Cytochalasin B	Unknown

Table 2 Synchrony procedures.

Type	Natural (N) or Induced (I)	Yield	Agency	Synchronizing action
Physical	N	Low	Gravity sedimentation	Small cells sink more slowly
	N	High	Zonal centrifugation	Centrifuge-assisted g, allows continuous collection
	N	Low	Drop-off (gravity)	Young, newly-divided cells fall off
	N	Low-Med	Shake-off (low Ca^{2+})	Mitotic cells selected
	I	High	Heat shocks	Works on division protein in protozoa
Chemical	I	Med	N$_2$O under pressure	Mitotic arrest augmented (shake-off)
	I	Low-Med	Colcemid/ Vinblastine	Mitotic arrest augmented (shake-off)
	I	Med-High	Methotrexate	G$_1$/S block, later released can be applied in two consecutive cycles
	I	Med-High	Excess thymidine	
	I	Med	Serum deprivation	G$_1$/S block, later released

(e.g. G_2/M boundary, see section 3.5) are held up. A great deal of cell cycle mapping has been done in this manner.

5.6   Synchrony

This is a most useful aid to analysis. In some fortunate cases, life provides *natural synchrony*, e.g. in early divisions of eggs, or nuclear division in *Physarum polycephalum*. Usually, however, chemical and physical constraints have to be used to hold cells at some point in the cycle (Table 2), i.e. *induced synchrony*.

Ideally, a synchronizing procedure should not disturb subsequent cell behaviour, but cells held for any length of time at some point will show *unbalanced growth*, in which cytoplasmic functions proceed while nuclear activity (the usual target of restraint) is checked. Some of the best procedures involve selection of cells by size or as mitoses (Fig. 5–4).

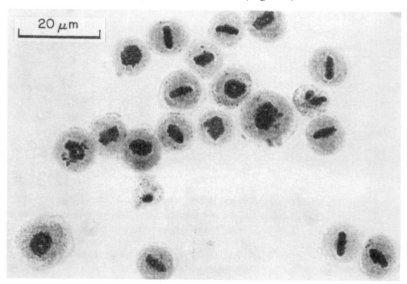

Fig. 5–4   Metaphase 'shake-off' synchrony of HeLa cells. A monolayer culture of asynchronously growing cells are washed with fresh medium and a small amount of new medium added to the cell layer, which is then shaken briskly. After repeating this procedure several times, almost pure metaphases are removed in the medium because they are the easiest to dislodge.

5.7   Fusion and the cell cycle

Fusion is the process of amalgamating two or more cells by means of attenuated virus particles or chemicals such as lysolethicin. Figure 4–4 illustrates the result of fusing an interphase cell with several mitotic cells, revealing the former's chromosomes and their own particular cycle. Much fusion work has aimed at answering questions such as whether a G_2 nucleus can be forced back into DNA replication by fusion with S-phase cells (RINGERTZ and SAVAGE, 1976). This does not occur but a G_1 cell fused in this way enters S more quickly.

5.8   Temperature-sensitive mutants

Reference was made in section 3.3 to the production of temperature-sensitive mutants. They are particularly useful 'tools' because they can perform a given function normally at one temperature (*permissive*), but fail to do so at a slightly higher temperature (*non-permissive*). It sometimes happens that mutants arise which can perform a given function at temperatures slightly above or below (e.g. 34 or 40°C compared with 37°C for mammalian cells) the optimum temperature for the wild type. Shifting cells to 'normal' temperature results in an arrest of this function at the appropriate termination point (p. 23).

Each temperature-sensitive mutant is usually the result of a single base alteration in the DNA of a gene – a *point-mutation*. This is one possible reason for their having a high rate of back-mutation or reversion. Mutants may be found for almost every cell function, although some genes are far more likely to be altered than others through treatment with such mutagens as MNNG (p. 23). Careful selection and analysis of them has been possible in many species including division yeast, brewers' yeast, bacteria and even mammalian cells. They have become instrumental in allowing the biologist to map by conventional genetics the gene sequences of chromosomes in these species. Their potential remains enormous and has become of added significance in giving information on gene position, which is of considerable importance in the rapidly developing field of biotechnology.

6 Hypotheses of Growth and Division Control

6.1 Introduction

Where mechanisms are obscure, ideas abound. But hypotheses are the starting points of scientific investigation, and good ones increase the chances of successful research. Before exploring some of these ideas, it is worth being refreshed about the main problems in this field:

(*i*)   What decides whether a cell will grow?

(*ii*) Is cell size related to chromosome number or gene dosage?

(*iii*) How does a dividing cell achieve bipolarization?

(*iv*) What precipitates events such as chromosome condensation or cytokinesis?

(*v*)   What generates the force to move the chromosomes to the poles?

6.2 The role of centrioles and cell centres

The word centriole is almost synonymous with division for some biologists. If cell functions were co-ordinated by something like a silicon chip, then this fascinating but extremely small organelle (Fig. 6-1) would be an obvious

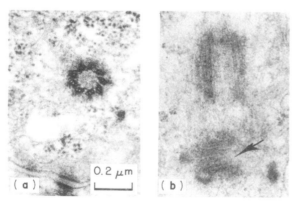

Fig. 6–1   Centrioles seen in T. S. (**a**) and L. S. (**b**). Oblique section of partner centriole (arrowed).

candidate for the task. For almost a century its behaviour strongly suggested that it polarised the spindle and controlled mitosis. Today its role is not so clear. For one thing, plant and other cells may have no centrioles. Furthermore recent laser-beam irradiation of the centriole has had no effect on a division in progress.

Before this venerable hypothesis based on the centriole is dismissed some further points must be noted. It has become obvious that the *pericentriolar matrix* is important and cannot stand laser beam damage. The dividing cell has two centres (Fig. 6–2) which derive from the original cell centre, but in the meantime are called *mitotic centres*. They are classed as microtubule organizing centres because they are capable of both assembling and disassembling microtubules. It is not certain whether the pericentriolar matrix is the epicentre of the mitotic centres, or how they arise. But perpetuation of the mitotic centres whatever these may be is important for all cell proliferation. As the mitotic centres separate to the poles (Fig. 6–2), the spindle develops between them. Cytokinesis ensures that each daughter cell has a new centre.

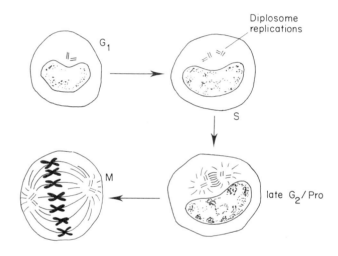

Fig. 6–2   Behaviour of cell centres during passage of cell into mitosis.

Spatial changes in dividing cells are amongst the most difficult to explain. Because the behaviour of microtubules in spindle formation is so prominent, one theory suggests that the events are controlled by a series of waves of Ca^{2+} sequestration and release through calmodulin (leaving unanswered the question of what regulates calmodulin). Release of Ca^{2+} prevents microtubules assembling and existing ones fall slowly apart. If this is followed in time by sequestration of Ca^{2+}, the cytoplasm develops a new superstructure of microtubules. The way in which these waves of Ca^{2+} can trigger different events is suggested in Fig. 6–3.

It is proposed that, first, calmodulin present in the cell centres releases Ca^{2+}, preventing tubules forming within this region. This Ca^{2+}-rich zone migrates outwards and allows the cell centres to move apart. The next stage is an increase in calmodulin activity with a consequent sequestration of free Ca^{2+} around the

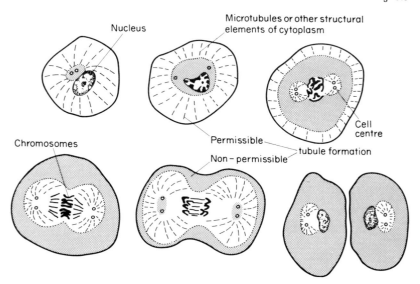

Fig. 6–3 Model for mitosis based on calmodulin activity causing waves of Ca^{2+} sequestration (clear zones) and release (stippled zones) in dividing sea urchin eggs. (After Harris, P. (1978) in *Cell Cycle Regulation*, eds. Jeter, J.R., Cameron, I.L., Padilla, G.M. and Zimmerman, A. M. Academic Press, New York, pp 75–104; with permission.)

(now clearly duplicated) cell centres, starting a wave of assembly of microtubules which, by their interaction, result in the formation of the mitotic spindle. As time passes, this wave in turn moves outwards causing tubules to interact with the cell membrane and cytokinesis to take place. At the cell centres, calmodulin once again causes the release of free Ca^{2+}, leading to a wave of disassembly and the rounding up of the daughter cells at telophase. After the sequence of events shown diagrammatically in Fig. 6–3, a fresh wave of Ca^{2+} sequestration usually occurs, allowing a cytoskeleton to appear and give a more differentiated shape to the cell. The hypothesis relies on intracellular Ca^{2+} concentration being inversely related to calmodulin concentration (activity?), and on direct observations of events in the dividing cell. Indeed, it specifically pertains to the behaviour exhibited by a favourite subject of biologists, the sea urchin egg. Some cells show far less re-organization during division. For example, in *Physarum polycephalum* and in yeast cells, segregation of the chromosomes at division can take place entirely within the nucleus. Even in mammalian cells, some remain relatively flat and differentiated during division, which suggests that the role of calmodulin and Ca^{2+} concentration may play only one part in a bigger programme of events responsible for dividing the cell.

6.3 Reservoirs, size and ratios

One simple suggestion for division control is that cells build up some component, say ATP, RNA or tubulin, to the point where enough is available (assuming everything else is in order) to set off mitosis. This presupposes that mechanisms exist which prevent cells entering mitosis without their reservoirs or complements being filled. Taken to a more logical end, it is when all substances reach their critical level that a cell divides, i.e. essentially its overall size. Size alone is not a controlling mechanism as can be shown by a simple example. If a bigger than average cell divides, its offspring will usually go through a quicker cycle than progeny from a smaller cell, and therefore both may return to a more uniform size before the next division. Something beyond size alone is involved. Nevertheless recent findings have shed new light on the role of size.

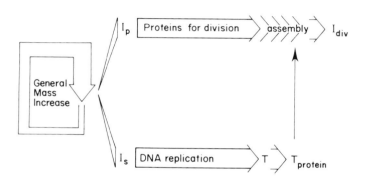

Fig. 6–4 A model for the co-ordination between DNA and division which assumes two pathways as well as general mass increase. Both DNA and proteins for division are set off (I_s and I_p respectively) in cycling bacteria, until at the end of S a special termination protein (T protein) is formed which instructs division proteins to assemble and initiate division (I_{div}). (After Jones, N. C. and Donachie, W. D. (1974). *Nature*, **252**, 252, with permission.)

No cell can divide below a certain critical mass, which will be different for each cell type. Most cells, however, divide at a size which is probably greater than the critical mass. In the *wee* mutants of *Schizosaccharomyces pombe*, which divides at about half the mass of the wild-type, the effect of size on the initiation of chromosome replication (S phase) and of nuclear division has recently been investigated. Although they behave normally at 25°C, their control over nuclear division operates at a much smaller critical size than in the wild-type when these temperature-sensitive cells are at 35°C. Initiation of S phase is not significantly affected, nor is generation time. Wild-type cells have a much larger mass immediately after division, which allows them to enter a new S phase quickly. The much smaller mutants, however, must go at least 30% of the way round their next cycle to reach the critical mass to enter S phase. Once this is accomplished, they can quickly move into division at half the mass of the controls. The conclusion is that at least two separate controls operate in normal cells, but the results do not prove that size *per se* is the crucial factor rather than some more pertinent regulatory mechanism, of which size is a crude index.

If size control is not acceptable because of lack of tangible mechanisms, then ratios could be important, such as the ratio of DNA to cytoplasm or membrane phospholopid to nuclear RNA. Such ratios could create a kind of chemical dialogue which could hold in check progression in the cycle until the right balance is achieved. As soon as this happens, certain events occur and the ratio is set down again. Systems like this operate on the oscillator-repressor basis. If ratios are indeed important in growth control, the need for separate initiator and suppressor substances would be greatly weakened.

Two theories of division rest heavily on the production of critical amounts of proteins for division, probably synthesized from S-phase onwards. When DNA is completely replicated, some signal is given in G_2 for the 'division proteins' to assemble and play their role in initiating mitosis. The model for bacteria is shown in Fig. 6–4; it could hold as well for most eukaryotic cells. The other theory is similar but dwells on the need to assemble division proteins in the cell. These are labile until they become consolidated or stabilized, which allows division to occur. When in their labile form, these structures are vulnerable to shocks (heat, chemical inhibitors) which break them down and cause division delays (Fig. 6–5).

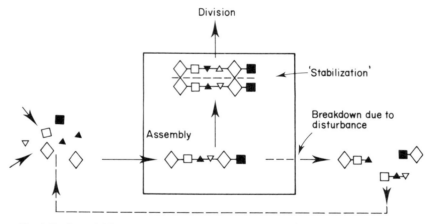

Fig. 6–5 A model for control of entry of cells into division. Assemblies of the various protein components (different symbols) must be made and stabilized before division can occur.

6.4   Programmes for division

It would be unproductive to enter into more than a few details about programmes. The easiest to imagine is that growing cells follow a set sequence of events which is read off the genome in linear fashion (the linear translation hypothesis). Alternatively other programmes might exist in which, instead of key events following in a linear sequence, several run parallel to bring eventually a cell to division by the co-ordination of their separate functions (parallel pathway model). The events which are depicted on such pathways generally involve the synthesis of some molecules. One important point to remember is that the time of *synthesis* is one thing, but the time of *utilization* of the product may be quite a different and more critical matter.

6.5 Transition probability hypothesis

Deterministic models (sections 6.3 and 6.4) have been questioned on the grounds that large variability in cell generation time between successive divisions are not easily explained. So Smith and Martin (in PADILLA *et al.*, 1974) proposed that after division, cells stay in an 'A' phase; they only move on at the command of some signal which acts quite randomly. This sends them into 'B' phase, which involves the set part of G_1 leading to S and so on through to the next interphase. The probability of moving through this transitional point in A varies with conditions, hence the name of transition or random probability hypothesis. A population of cells will show the type of kinetics of entry into division seen in Fig. 6–6, in the α-curves. In this figure, a β-curve is also plotted.

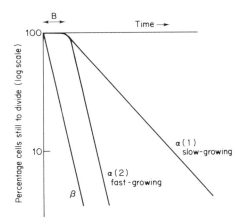

Fig. 6–6 Plots for 2 different cell lines (1) and (2), both taking time *B* to complete their deterministic functions, but having different rates of passing through *A*. 'α' plots give an indication of cycle time, and its variable *A* and fixed *B* components. In particular, the semi-log plots demonstrate a truly random, first-order (exponential) decline of cells moving out of *A* phase.

The β-plot, based on α(2) data, compares the difference in cycle time between each pair of daughter cells in the subsequent cycle. If cycle time was genetically inherited or fixed, the difference would be negligible, and the curve run down the ordinate. The β-plot, however, is parallel to the slope of α(2) and therefore siblings show the usual randomness in moving on through *A*.

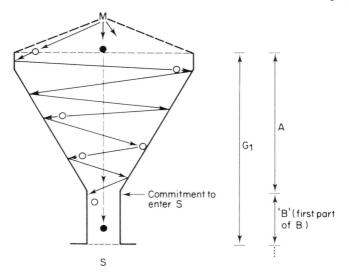

Fig. 6-7 The hopper model to explain how G_1 may be naturally variable. A and B refer to probabilistic and deterministic periods respectively.

This is the *difference* in time between two siblings to enter their next division. It shows the real randomness of the event in A phase. In an attempt to account for variation in cycle time in a less abstract and mathematical form a model based on a 'hopper' principle is invoked to explain the nature of the randomness of the G_1 duration times during which a powerful growth and cell cycle control is exerted. Figure 6-7 illustrates the principle. Cells take different paths through the 'hopper', that is, spend different times, randomly determined, in the 'A' phase of G_1 before commitment to DNA synthesis in S. Some cells may fail to pass through the 'hopper' and they remain, more-or-less permanently in early G_1, a state called quiescence referred to as G_0.

7 Regulation of Growth and Division

7.1 Cellular nutrition

Maintenance and growth are critically dependent upon an adequate supply of nutrients (Chapter 2), and any substance which alters transmembrane availability, e.g. insulin or thyroxine, could exert a regulatory influence. Where intake is not limiting, intrinsic mechanisms must take over full control. Incessant growth, as in cancer, has been linked to both environmental factors causing permanent changes in the assimilation of nutrients and to genetic changes which erase the intracellular controls.

7.2 Contact and cell density

Cells usually grow into open spaces, and in a wound they quickly close it up. Signals pass between cells, often at distance from each other, which can be mutually inhibiting; others grow until they are crammed together (Fig. 7–1).

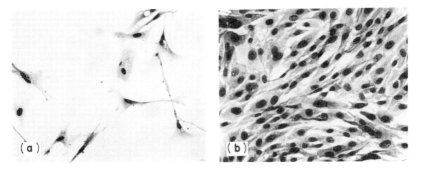

Fig. 7–1 (a) A sparse culture of mouse fibroblast cells which will become inhibited after little more growth. (b) Another cell line grows to confluency.

Malignant cells go further and grow over one another. Contact between cells tends to stop both movement and cell division. Their membranes must interact to send signals to the nucleus which regulates division, a response which can be modified by environmental factors and hormones. Loosening contacts with trypsin will release cells from restraint, and they may go through one or two divisions. It is possible that some of the stimuli which induce new mitoses may act through adenyl cyclase in the membrane. This converts ATP into cyclic-

AMP (cAMP) in the cell. One theory of cell regulation – the Yin-Yang hypothesis – maintains that the cAMP: cGMP ratio in the cell influences its proliferative behaviour (HOGAN and SHIELDS, 1974).

7.3 Membrane phenomena

Definite changes in the transmembrane potential occur in proliferating cells. Cells with potentials of up to 100 mV will show a fall to 20 mV or lower at division. Cancer cells have lowered potentials. These changes reflect alterations in ionic balances across the membranes and are presumably a result of intracellular alterations.

Another property of membranes affecting, or affected by, growth is their fluidity. Membranes containing many saturated fatty acids are more rigid and less adaptable to, for example, temperature shifts than those containing mainly consaturated fatty acids. This may also affect permeability and other properties mentioned above. The membrane remains one of the most studied parts of the cell in relation to proliferation, because it responds first to any environmental challenge.

7.4 Chalones

Tissues may produce and accumulate substances which dampen down their own activity. Oddly, 'chalones' – as these substances are called – are tissue-specific but not species-specific in vertebrates. If an organ is damaged, the total chalone it is dispersing will be cut down, restraint will be removed and the injured tissue regenerated. While the chalone theory concept seems logical, it has only been demonstrated satisfactory in a few tissues and as a general mechanism remains equivocal.

7.5 Regeneration: hypertrophy and hyperplasia

Regeneration is an extensively studied area of cell growth, requiring a whole book to discuss. Growth induced by a functional load through loss or injury can result in *hypertrophy*, increase in cell size, or *hyperplasia*, increase in cell number. Both usually occur, but hypertrophy generally ends when numbers restore the loss. Although studied in severe cases, removal of half a *Hydra* or two-thirds of a rat liver, regeneration is an exaggeration of a constant process of cell *turnover*. Populations lacking turnover, e.g. brain neurones, also lack regenerative potential. In cell populations showing stimulation into division after injury, G_1 or G_0 (quiescent) cells are brought back into cycle. Some tissues may have arrested G_2 cells which can quickly move into M.

7.6 Cancer

A short overview of cancer can be found in HARRIS, 1970. This condition has attracted great attention not entirely for altruistic reasons, but because the pathology associated with the loss of cell regulation should teach us something about the normal control. While cancer is undoubtedly of many different origins, it is noteworthy that in no case can a clear pathway of action be described between the causative agent and the provoking of the cell into repeated divisions. The reason is probably because there may be hundreds of influencing factors along this pathway which complicate matters. The important fact about carcinogenesis is that, unlike regeneration, when the stimulus stops the cells nevertheless go on proliferating. Cancer cells do not necessarily show abnormal behaviour other than in incessant proliferation; they show the activity of rapidly growing and dividing cells of their normal counterparts. By persisting in this activity they may end up infiltrating and spreading into undesirable places. Small differences in growth potential of a population, or in drop-out (see section 7.7) could be enough to produce tumours.

It is widely assumed that cancer must be due to alterations in the genetic constitution of cells, involving mutation in the genomic information (DNA) in the nucleus of the cell. One problem is that careful pathological examination of pre-cancerous tissues often suggests a *field effect*, as if many cells are apparently responding in a similar manner at the same time in response to a carcinogenic stimulus rather than cancer coming from a clone of mutated or transformed cells. Tumours (malignant tumours are commonly referred to as cancers) should be mostly homogeneous collections of cells if a single cell mutation is involved. It also follows that if mutations of genes responsible for controlling cell growth and division are the cause of cancer, all carcinogens must be mutagenic, or at least potentiators of spontaneous mutation rates. Until recently, this had not always appeared to be the case because tests of mutagenicity did not simulate with sufficient accuracy the metabolic conversions that chemical carcinogens can undergo within living tissues. It now appears that many carcinogenic substances in their 'native' form are relatively ineffective as mutagens until they have been activated or chemically converted into more reactive species within the body tissues. To preserve meat, nitrites are sometimes added, which may be non-carcinogenic. Within the gut and tissues, nitrosamines can be formed from them, which are potentially far more hazardous. Nitrites turn out to be *proximate* not *ultimate*, carcinogens.

The mutation theory of cancer, however, cannot be taken as proven. Cancer may be due to other disturbances in cells. One is the *oncogene hypothesis*, which suggests that the genome of many cells may contain, or be infected with, potentially dangerous virus-like pieces of DNA. Activation of these sequences by carcinogens results in the oncogene switching off or by-passing the growth regulatory genes of the host cells, causing the latter to proliferate incessantly. But unrestrained division necessitates a constant supply of nutrients. Many theories of cancer have focussed on the possibility that the plasmamembrane of cells is altered in transformed cells (cells taking on malignant characteristics). This could make them more permeable to nutrients whose entry under normal circumstances would be carefully regulated. Membranes of some tumour cells show an unmasking of normally hidden antigenic determinants. Or they have much lower transmembrane potentials than their untransformed counterparts, which could account for their changed permeability and mobility. Tumour cells become particularly lethal when they are highly mobile, invading other tissues and causing *metastases*. Changed membranes could result in the loss of contact inhibition not only of movement but of proliferation.

A perennial problem for the cell biologist exploring the differences between normal and malignant cells is to separate cause from effect, and this has undoubtedly been responsible for the relatively slow progress in cancer. It has also been suggested that cancer need not be a disease of individual cells, but a result of aberrant behaviour occurring from time to time by groups of cells.

A simple, although not exact, parallel is disturbances in human society. This is often caused by criminals tending to destroy the harmony of the system, but who are nevertheless difficult to find because they look like other human beings except when they are indulging in their anti-social behaviour. It is just as difficult to brand a cell as malignant as it is to call a man a criminal, for it is often a consequence of the company it keeps and the prevailing state of affairs which may make it lose the control which normally keeps it in its place.

7.7 Cell death in relation to proliferation

Although the processes of growth, cell renewal and so on have been discussed at length, cells and organisms must be seen in terms of populations. A population in balance, where division compensates for death, can quickly change to growth or atrophy. Ways in which this can be achieved are shown in Fig. 7–2, but mechanisms which allow this regulated change are poorly

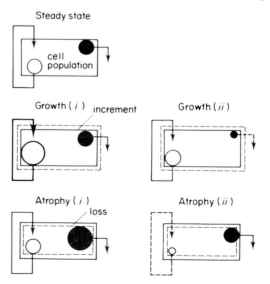

Fig. 7–2 Diagrams to show how alteration in either growth fraction (○) or drop-out fraction (●) can alter the overall size of a cell population.

understood. Concentrating on cell drop-out, in some cases this will be due simply to the loss of cells from the collective population, as in the shedding of mucosal epithelial cells from the tips of the villi in the gut. A second method is through sudden and quite dramatic autolysis or osmotic lysis. The former is probably due to lysosomes (Fig. 1–3) becoming very active. Another type of cell death occurs through cells undergoing progressive condensation of both their cytoplasm and nucleus to give a shrunken or apoptotic cell which may eventually be engulfed by phagocytic cells. What process determines that some cells should die, often by some kind of programming, remains unknown.

Further Reading and References

AHERNE, W. A., CAMPLEJOHN, R. S. and WRIGHT, N. A. (1977). *An Introduction to Cell Population Kinetics*. Edward Arnold, London.

CARTER, B. L. A. (1970). Unit 7; S-321 Course, Open University Press, Milton Keynes.

CAMERON, I. L. and THRASHER, J. D. (1971). *Cellular and Molecular Renewal in the Mammalian Body*. Academic Press, New York.

EDWARDS, C. (1981). *The Microbial Cell Cycle*. Nelson, Walton-on-Thames.

DAWKINS, R. (1978). *The Selfish Gene*. Paladin Books, Granada Publishings Ltd., London.

DIRKSEN, E. R., PRESCOTT, D. M. and FOX, C. F. (Editors) (1978). *Cell reproduction: in honor of Daniel Mazia*. Academic Press, New York. (Many short papers by eminent investigators)

HANAWALT, P. C. (1980). *Molecules to Living Cells*. Readings from *Scientific American*. Freeman, Oxford.

HARRIS, R. J. C. (1970). *Cancer*. Penguin Books, Harmondsworth.

HOGAN, B. and SHIELDS, R. (1974). *Yin-Yang Hypothesis of Growth Control*. New Scientist 9 May, p 323.

JOHN, B. and LEWIS, K. R. (1972). *Somatic Cell Division*. Oxford Biology Readers No. 26. Oxford University Press, Oxford.

LITTLE, M., PAWELETZ, N., PETZELT, C., PONSTINGL, H., SCHROETER, D. and ZIMMERMANN, H.-P. (1977). *Mitosis: facts and questions*. Springer-Verlag, Berlin.

LLOYD, D., POOLE, R. K. and EDWARDS, S. W. (1982). *The Cell Division Cycle: temporal organization and control of cellular growth and reproduction*. Academic Press, London.

LUYKX, P. (1970). *Cellular Mechanisms of Chromosome Distribution*. Academic Press, New York.

MITCHISON, J. M. (1971). *The Biology of the Cell Cycle*. Cambridge University Press, London.

NICKLAS, B., BRINKLEY, B. R., PEPPER, D. A., KUBAI, D. F. and RICKARDS, G. K. (1979). Electron microscopy of spermatocytes previously studied in life: methods and some observations on micromanipulated chromosomes. *J. Cell Sci.*, **35**, 87–104.

PADILLA, G. M., CAMERON, I. L. and ZIMMERMAN, A. (1974). *Cell Cycle Controls*. Academic Press, New York. (Excellent papers by Zeuthen; Smith and Martin; Mitchison; Howells and many others)

PRESCOTT, D. M. (1976). *Reproduction of Eucaryotic Cells*. Academic Press, New York.

RINGERTZ, N. R. and SAVAGE, R. E. (1976). *Cell Hybrids*. Academic Press, New York.

Index